CANNEL COAL OIL DAYS

CANNEL COAL OIL DAYS

BY

THEOPHILE MAHER

EDITED BY

EDWARD WATTS

WEST VIRGINIA UNIVERSITY PRESS
MORGANTOWN

First edition published 2021 by West Virginia University Press
Printed in the United States of America

ISBN 978-1-952271-11-3 (cloth) / 978-1-952271-12-0 (paperback) /
978-1-952271-13-7 (ebook)

Library of Congress Cataloging-in-Publication Data
Names: Maher, Theophile, author. | Watts, Edward, editor.
Title: Cannel coal oil days : a novel / by Theophile Maher ; edited by Edward Watts.
Description: First edition. | Morgantown : West Virginia University Press, 2021. Includes bibliographical references.
Identifiers: LCCN 2021009485 | ISBN 9781952271113 (cloth) | ISBN 9781952271120 (paperback) | ISBN 9781952271137 (ebook)
Classification: LCC PS2359.M6456 C36 2021 | DDC 813/.4—dc23
LC record available at https://lccn.loc.gov/2021009485

Cover design by Than Saffel / WVU Press

Contents

Acknowledgments

When I was a child, two enormous white pines dominated the yard in front of my father's family's longtime cottage across the road from Lake Michigan just south of Saugatuck, Michigan. We called the trees Theophile and Sarah, after my great-great-grandparents who first bought the land around the turn of the twentieth century as an escape from the summer heat of Chicago, where they settled after the Civil War. Until 1910, the family camped on the property. Then Theophile and Sarah Maher's son George, a successful architect, built a cottage on a rise for his recently widowed sister, my great-grandmother Mary Maher Hooker. Theophile and Sarah, the trees, lasted about a hundred years and died surrounded by dozens of their children, most of which still stand.

As the fifth in the family line to own the land, I write this in the summer of 2020 on the porch of that cottage, shaded by those pines. By bringing Theophile's fictionalized account of his and Sarah's years in western Virginia into print, I hope to repay some of the debt I owe my forebears. The handwritten manuscript for this book was passed among different descendants of Theophile and Sarah until the 1970s when Elizabeth Wyld, descended from the girl Harriet here, gave it to my mother, Emily Stipes Watts, the wife of Mary Maher Hooker's grandson, and a professor of English at the University of Illinois. While I had been aware of the text's existence, I had not seen it until after her death when I found it in the family archive she so diligently established. Fascinated, I read it and started it on its path to publication.

Along the way, I incurred a number of debts. First, thanks to Dean Rehberger, director of MATRIX at Michigan State University, not only for decades of fellowship and collegiality, but also for having Callie Hamm and Anna Sherry use Transkribus, a developmental program for digitizing texts, to translate the handwritten pages into a functional text I could edit. At West Virginia University, I thank professor of English Timothy Sweet for helpful advice; at the Press, Derek Krissoff for his patient guidance through the peer review process; and of course, the anonymous readers. While I had edited nineteenth-century manuscripts before, their input made this project better in many ways. From Chicago, historian Kathy Cummings offered her expertise on George Maher and elements of my family history about which I knew very little.

Finally, I thank those who kept Theophile's and Sarah's family and land through the turbulent twentieth century: Mary Maher Hooker, Tony and Margaret (Hooker) Janata, and Robert and Emily Watts. My late brothers (Ben and Tom) and my cousins (Robert Dawson, Amy Zoller, Beth McIlhaney, Richard Watts, and Wendy Zemanski) and their children have also benefited from this legacy. The family Theophile and Sarah began has traveled far from the hills of the Kanawha valley and Mill Creek but their story reflects the values and courage to which we might all aspire. In the end, I thank Theophile for his tale as our inheritance.

Fig. 1. Portrait of Theophile Maher (editor's family collection)

INTRODUCTION

CANNEL COAL OIL AND THE CREOLE COSMOPOLITAN

Edward Watts

Theophile Maher (1815–1902) was the son of a French Huguenot, Edouard St. Hilaire Maher, who migrated to Haiti (San Domingue) to escape the Reign of Terror during the French Revolution. Subsequently, he fled Haiti to escape Touissant L'Ouverture's Rebellion, first to Norfolk, Virginia, and then to Philadelphia. Once there, Edouard was prominent enough to merit a portrait by Gilbert Stuart. Theophile's mother, Sophie Cecile Clavier, was likewise a Creole refugee from Haiti whose family had also settled in the City of Brotherly Love. As a young man, Theophile trained as a chemical engineer at the Sorbonne in Paris, where he knew Alexandre Dumas *père* and the founders of the newspaper *Le Figaro*. Later, he unsuccessfully mined for gold in California and soon returned east to start over. At some point, he converted to Methodism, likely as part of the Second Great Awakening, and his Christian values from then on strongly affected his work and life.

In 1850, he married Sarah Landis of Lancaster, Pennsylvania, kinswoman to Kennesaw Mountain Landis, later the well-known commissioner of Major League Baseball. After their marriage, Theophile and Sarah relocated first to Chicago, where Theophile worked as chemical engineer and refined new techniques in oil distillation before a fire destroyed his laboratory. From there, he settled in Cincinnati and later Kanawha County, Virginia, ten miles up the Elk River from Charleston, in what is now West Virginia, before returning to Philadelphia after the Civil War. Later Theophile and Sarah relocated to Baltimore and, after stops in Indiana and Kentucky, finally to Wilmette, Illinois, a suburb of Chicago, to be near family, including their son George Maher—a prominent architect and early associate of Frank Lloyd Wright.[1]

Yet among all his various activities—a trek embodying the transformation of the United States during the nineteenth century from collection of divergent former colonies to industrial nation—when he decided to tell his story, Maher's time in the coalfields of western Virginia, and his involvement there in the opening days of the Civil War, most strongly struck his imagination. More precisely, his time implementing new processes for distilling oil more cleanly from cannel coal in a hamlet named Mill Creek off the Elk River's south bank ten miles upstream from Charleston (not the better-known Mill Creek in Jackson County) and then organizing local Union efforts at the start of the Civil War between 1859 and 1861 inspired the literary project he engaged as an older man. In 1887, using lined stenographic notebooks, he wrote a semiautobiographical novel, *Cannel Coal Oil Days*. This fictionalized memoir, reproduced here for the first time, has been taken directly from his handwritten manuscript. In addition, his notes suggest that this was only the beginning of a larger text he had drafted, but not completed. Those additional materials cannot be found.

As described below, the text itself fictionalizes Maher's time in western Virginia as it became West Virginia. It begins with his arrival in Charleston, then a village Maher refers to as "the Kanawha Court House" and traces his trip to Mill Creek. There he works as mining engineer for a Kentucky-based company set on improving the efficiency and production of its mining and distilling operations. The story continues through his entry into and interaction with the mining communities upstream, both white and enslaved. It minutely details how labor, transportation, education, religion, politics, economics, race, and, at the center of it all, mining operated in the Kanawha valley in the mid-nineteenth century. The novel draws to a close with the arrival of a Confederate force into the staunchly Union stronghold as it comes up the Elk River to Mill Creek. Mr. Mark—a thinly disguised version of Maher himself—and his neighbors, all Lincoln Republicans, participate in and support the secession of West Virginia from Virginia, and the nascent state's advocacy of both the abolition of slavery and loyalty to the Union.

In the final scene, the hero, a stringent abolitionist and environmentally conscientious mining engineer, is pursued by the Confederates for organizing the local Union militia. At novel's end, Mark surrenders to avert a violent confrontation between the mountaineers and the Rebel troops. Although the surviving and incomplete manuscript ends abruptly, here appendix A reproduces a 1901 newspaper story describing Maher's own recollection of his brief captivity and later rescue by an Ohio regiment. More artifact than art, *Cannel Coal Oil Days* brings to life, in a direct and straightforward manner with a number of subplots and undercurrents, a moment in West Virginia history largely overlooked in the materials available: life in antebellum western Virginia's mining country and the coming of the war. Some explanation of the story itself and its late nineteenth-century contexts—such as mining history, Maher's use of local color and progressivism, slavery and mining, and the coming of war to western Virginia—will bring the book and the changing of mid-nineteenth-century Appalachia into the twenty-first century.

A Story of Transformation

The division marked by the colon in the original title captures the novel's two main narrative strains in a way that mirrors the region's own transition from Appalachian backwater to site of interregional conflict: *Cannel Coal Oil Days: A Reminiscence of the War of Rebellion*. While the colon might imply two separate stories—one of mining and one of the run-up to war—in fact, each is a component in the larger story of how Mr. Mark brings reform and modernity to the region. Whether he is improving the efficiency of cannel coal oil production, bringing the locals back to the church, cleaning up polluted streams, insisting the company pay leased slaves, stumping (literally) for West Virginia to secede from Virginia, or organizing the Union militia, Mark's actions are informed by his dedication to a version of progressivism based in his Christian faith. As much as the novel depicts events from 1859 to 1861, it also reflects the postbellum moment of its composition in 1887.

The narrative's first sections describe the entry of Mark into the Elk River region as the representative of a cannel coal oil company whose owners all reside in Kentucky and operate out of Lexington. The company purchased two thousand acres, beneath which is cannel coal, a soft type of coal whose distilled oil was marketed for lighting in the years between the overfishing of the sperm whale and the development of electric lighting. The company has put in charge two unqualified men, and production and profits have been disappointing. Cannel coal's mining and distillation were initially very messy and arduous, though, in the novel, Mark develops and puts in place methods for both improving production and lessening the environmental impact of its production.[2] Most of the miners and laborers are slaves leased from enslavers to the east, in "Old Virginia." Once he arrives and understands their maltreatment and how that contributes to their lassitude, Mark insists on paying them after he wrests control of the mining operation from Fieldbell and Reynolds, the previous incompetent and corrupt managers.[3] Treated by Mark with dignity, the slaves respond positively: the problem is not the slaves themselves but the institution itself. Abolition and antiracism characterize even the story's early mining-based sequences, before moving to the center near the end of the novel.

Fig. 2. Picture of Elk River, c. 1870 (West Virginia and Regional History Center, WVU Libraries)

Through its middle sequences, these themes are merged. The text follows the development of the company's coal business through the experiences of Colonel Albert, one of its primary investors, as he, his son, and Mark root out wasteful corruption, inefficiency, and the bad citizenship of the industry more generally. Along the way, Mark and the Colonel engage reforms that reflect, even now, ongoing concerns in the region and the nation: corporate integrity, racial conflict, and mining's history of environmental damage. Maher addresses both through the lens of a post-abolition progressive ideology reflective of his generation, one raised in the moment of the Second Great Awakening, a nationwide spiritual revival (1800–1840) that inspired Americans in the North to pursue not only abolition but also temperance, women's rights, and labor reform, among other causes, in the name of aligning the nation more closely with "simple Christian" values.[4]

In the novel's final third, as the war approaches, interregional conflicts redirect the narrative. Mark and his mountain-based neighbors demonstrate both their abolitionist and Unionist loyalties. At this point, the primary family among the enslaved population comes to be tightly entangled with Mark, his wife Sarah, and their children, as well as the other Unionists in the neighborhood. In particular, Mark interacts with the son John, who has purchased his freedom and

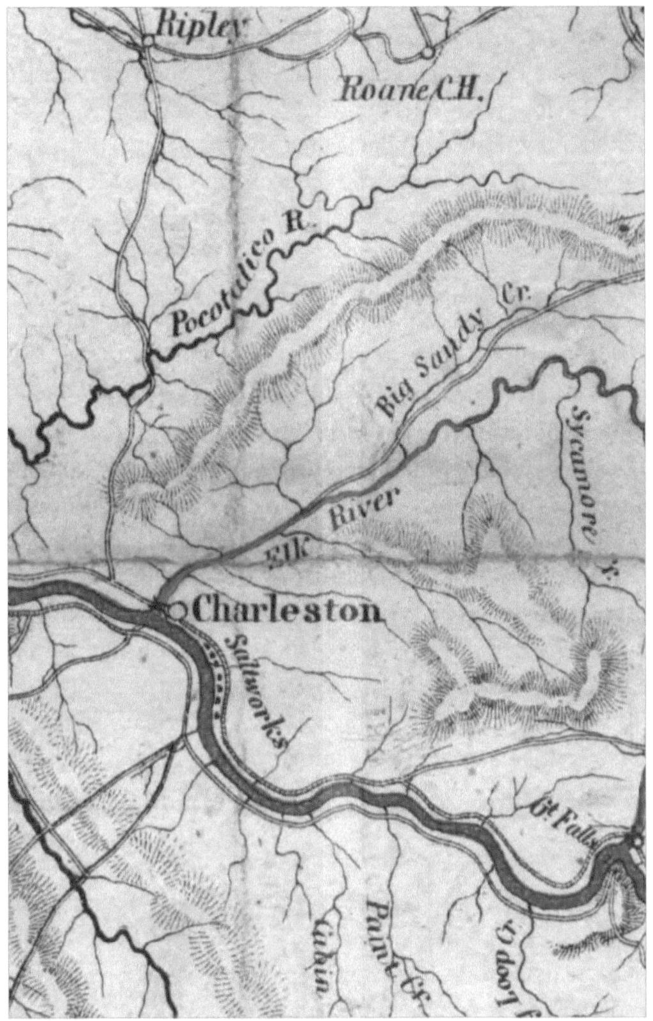

Fig. 3. A map of the Elk Valley from 1865. Mill Creek runs south from the Elk at the end of the marking (Library of Congress, Geography and Map Division)

is employed in Wheeling, where the proponents of seceding from the rest of Virginia have convened. John travels the region and dispenses information surreptitiously: slaves often learn of developments well before the white characters.

Mark trusts John with the militia's money to buy and deliver much-needed gunpowder, and John delivers. In fact, John and his family often embody values closest to Mark's own: family, faith, work, and service. Maher's consistent representation of the Black population as articulate, resourceful, brave, and intelligent simply ignores stereotypical representations. These characteristics are presented silently in the text as if, writing in the days of Jim Crow, Maher portrays the morality, humanity, and patriotism of Black Americans as too obvious to require further explanation.

The local white Appalachians are also well represented. While some are scurrilous types whose time is ended by Mark and the Colonel, of more interest are two strains of the mountain's free population. First are the teachers and leaders. The schoolteacher Mr. Clark and the Reverend Freenberry are both virtually penniless, but each devotes himself to the betterment of the mountain folks. Second is the mountain folk themselves. Early on, Mark explains how local settlers—"squatters"—had been cheated of the lands by wealthier landowners from eastern Virginia who forced them to pay rent on lands they had already cleared and farmed. Despite this, wholly absent are any long-established stereotypical representations of Appalachians as "white trash," even in their views of Black Americans.[5] Emblematic here are the Druker and Johnson families. Druker operates the steamship up and down the Elk River and is inspired by Mark to rededicate himself to a life of Christian virtue. Later, Druker bravely sneaks past the Confederate troops to bring crucial information up into Mill Creek. Johnson is Mark's chief miner and consistently treats the leased slaves with respect and integrity.

The women of Mill Creek, in particular Sarah, Mark's wife, as well as Mrs. Druker and Mrs. Johnson, both affirm and transcend Victorian strictures of passive femininity. While each performs the standard role by personifying domestic virtue, each also speaks her mind. Mrs. Johnson in particular confronts a Confederate soldier and convinces him to desert and return to Kentucky and his wife. As to the stand-in for Sarah Landis Maher, Mrs. Mark, the texts reproduced in the three appendices both confirm and complicate the historical accuracy of her role in one of the novel's final events: both the fictional and actual Sarahs sew a flag to support the referendum to secede from Virginia, to fly over the polling station in Charleston. As there is no Union flag in the Kanawha valley to be found, the Marks ask that John procure red, white, and blue silk from Wheeling, and Sarah and Mrs. Johnson expeditiously construct a flag around which Union and separatist forces can rally. The diverse Appalachian community Maher revives, however, remains based on the coal under the ground.

Cannel Coal and Coal Oil

The first and middle sections of the novel describe at length Mark's modernization of his company's cannel coal oil mining, distillation, and production processes. Cannel is very soft, malleable coal whose only current usage is handcrafted sculptures. However, in the mid-nineteenth century it was briefly highly profitable when distilled and sold as an oil that could function in lamps developed to burn once-abundant whale oil. In North America, cannel coal can be found especially throughout the Ohio valley, ranging from Appalachia through southern Indiana and Ohio and northern Kentucky in particular in an extended vein known as the Breckenridge Line.

When Mark arrives in Charleston from Cincinnati prior to his moving upstream to Mill Creek, he visits the company's distillery where the crude oil extracted from the coal at the mouth of the mine is refined to safe-burning oil for sale and shipment down the Kanawha River to the Ohio and markets downstream. As described by Henri Erni in 1865, coal oil was extracted from crushed coal first by being superheated in retorts (Mark favors clay ones). The resulting vapors were collected and piped into condensers where they were treated with various alkalis to clean and purify the distilled fluid. Mark's major innovations are at this stage, as he has developed a new way of cleaning the oil. Maher describes the condensers and stills used in these processes, as well as the resulting tar, ash, coke, and other by-products, many of which have despoiled the Appalachian environment. The coal ash left behind in the retorts was initially left as waste, but, as Maher notes, the resulting runoff from these substances was toxic, and the novel describes Mark's insistence on burning off the ash in safe containers, preventing both the poisoning of the streams in Mill Creek hollow and at the refinery in Charleston.[6]

When Mark arrives at the mine, he considers it his duty to find cleaner and more efficient means of refining the entire process. After his first year, the streams have been cleaned, the distillation machinery has been shifted to the mouth of the mine itself, and cleaner and better oil is being produced at a lower cost. Mark's changes—as well as his more honest accounting and management—are accomplished not only in the spirit of profit but also in the spirit of genuine scientific curiosity and concern for the beauty of the region and the health of its inhabitants. However, the cheapness of cannel and other coal oils, even in 1861, was becoming a moot point. From the start, Maher positions the discovery and improved production of petroleum and kerosene as constant threats to the commercial viability of cannel coal oil.

As to the business of cannel coal oil itself, in 1865, Abraham Gesner, whose father had pioneered the distillation of coal oil in the 1840s, published *A Practical Treatise on Coal, Petroleum, and Other Distilled Oils*, a general history and how-to for the industry as it had developed prior to the Civil War. Gesner begins by noting the potential of coal oil. Beginning in the 1840s, he comments, "The refining process was not so well understood at that time as at present, and the odor was not agreeable. The beauty of the light from it, however, was sufficient to gradually overcome the objection on the score of odor. Its supposed explosiveness was also urged against it by those interested in the camphene and burning fluids" (11). Despite these unfair and inaccurate aspersions cast by these rivals, Gesner notes that progress in coal oil's production methods "has been gradual . . . [and] carried on not by the labors of one mind, but of many, so as to render it difficult to discover to whom the greatest credit is due" (10). Given Mark's use of an alkali that he had developed while in Chicago to reduce the odor and purity of his cannel oil, it seems likely that Maher was among those responsible for this incremental progress. Gesner also credits advances in coal oil production with informing the development of crude oil distillation (12). Gesner prefaced these remarks by noting the seeming irreversible slide of coal oil into obsolescence: "The great cheapness of the oil procured by the distillation of petroleum has, however, almost caused the coal distillation to be suspended. It will only be resumed when the petroleum wells cease to yield sufficient oil for the various purposes to which it is now applied" (10). Of course, writing in 1865, Gesner could not know that Thomas Edison would soon make petroleum itself obsolete in the lighting market, even as the fuel awaited Henry Ford's exponential redeployment of it in internal combustion engines.

While Maher knew of the demise of coal oil for lighting as he wrote in 1887, the product to which Mark (and Maher) dedicated their labor is of secondary concern. Mark's primary interest in improving his techniques, and Maher's in creating the novel, transcended the coal oil industry. Instead, each was more concerned with the linkage of technological and economic advancement to social and moral progress, reflecting a greater concern for the well-being of the emergent industrial nation and, for Maher, the role of writing, specifically the popular postbellum mode of local color realism, in the work of faith-based social reform.

Local Color in Mill Creek

As he tells these stories, Maher writes in the vein of local color realism, especially in the fourth chapter, although his amateurism occasionally cannot be avoided.

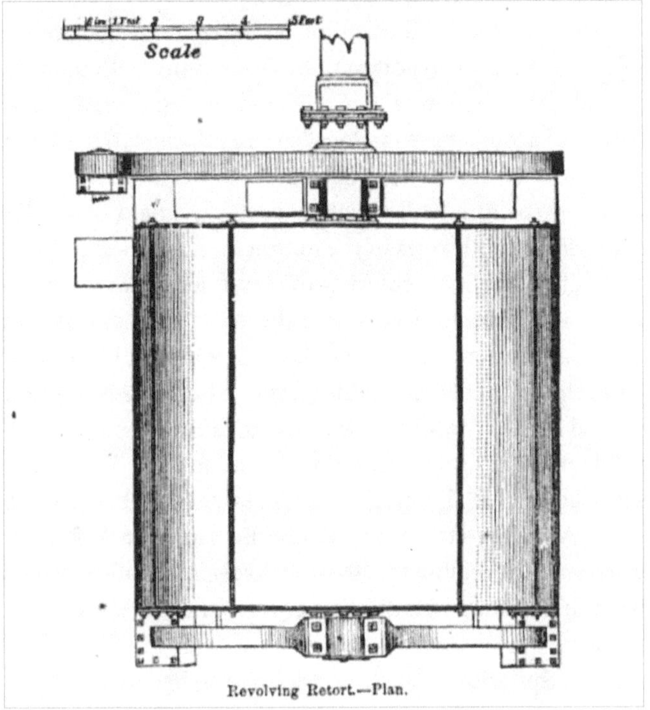

Fig. 4. Gesner's sketch of nineteenth-century coal retort (Abraham Gesner, *A Practical Treatise on Coal, Petroleum, and Other Distilled Oils*, 2nd ed. [New York: Bailliere Brothers, 1865])

As it emerged in the late nineteenth century, local color reflected the feelings and desires of a generation of Americans that had matured prior to large-scale industrialization and remembered back to when "America was bred in a cabin," as phrased by British traveler Morris Birkbeck in 1817. To that end, geographic setting likewise defines local color. Maher often evokes the wild topography of the Elk valley to recreate a beautiful natural setting to counterpoint the messy work of mining and politics, as when Mark discovers how many local streams have been contaminated by the extraction process, compelling him to work to restore their original clarity and purity. Along the same lines, as industrialization, telegraphs, railroads, large corporate entities, and wire services made each American locality more alike, local color reminded nostalgic readers of earlier place-specific moments, when communities were more self-contained.

In turn, realism's rejection of more romantic and speculative narratives also reflected a broader agenda based in local color's relocalization of national identity. Realism reminded the healing nation of its prewar identity. In this book, by reinventing antebellum Mill Creek as part of a chain of small yet similar rural communities on both sides of the mountainous terrain of the Ohio valley, Maher joined with other local colorists to recraft a common national memory of cooperation and community to heal the war's disruptions. For example, with industrialism came much starker class distinctions, but in describing the educated and refined Mark among his workers, both Black and white, Maher writes: "Day and night, [Mark] was with them, among them, one of them, and as he loved his work, so they loved theirs also and tried hard to get a good word from him about it at all times." In this passage, the northern cosmopolitan communes with working-class white southerners and Black slaves in a tableau of shared experience, reversing the divisions accelerated both by the war and by industrialism's accelerating stratification and separation of the nation's regions, races, and classes.

To fulfill this aspirational function, realism also had to depict the era's problems. As encouraged by William Dean Howells from his position as the editor of two prominent literary periodicals, realism imagined how fiction could bring to light these problems. In fact, one of the first and most important texts in the genre was likewise based in western Virginia: in *Life in the Iron Mills* (1861), Rebecca Harding Davis creates a horrific picture of the degrading industrial conditions at a coal distilling plant in Wheeling. However, instead of taking a transcendent approach to similar issues (as had Nathaniel Hawthorne or Herman Melville), realism in the hands of Davis and, I would add, Maher addressed national issues at the more scalable level of community, neighborhood, and family to inspire their readers not to contemplate these tragedies but to work toward repairing the nation through progressive moral and cultural reform.

In many ways, *Cannel Coal Oil Days* occurs on the many borders that defined nineteenth-century America: North and South, Black and white, modern and traditional, urban and rural, and middle and working classes. It exposes how these divergent presences might be reintegrated to heal the nation through the inclusive values of Christian progressivism—simple kindness and respect. The Civil War disrupted the development of what could have been at Mill Creek: a profitable, clean, and multiracial place for families to live and work together in peace. Though he was writing about 1860, Maher's message for his imagined 1887 readers, therefore, endured, as problems of race, region, and class persisted into the postbellum era.

Yet his experiences before he traveled to Mill Creek contributed to his authorial motivations. Howells's earliest poetry was published by William T. Coggeshall in his massive collection *The Poets and Poetry of the West* in 1860. Coggeshall worked at the tail end of antebellum Cincinnati's literary efflorescence. Having spent time in Cincinnati in the 1850s, and as a staunch Republican and abolitionist, Maher was likely familiar with these literary figures and shared their political and cultural values.[7] After the war, Howells moved to Boston to edit the *Atlantic*, and later to New York to *Harper's*. From his position atop the American literary establishment, Howells advocated and championed local color and realism. The fiction he championed was not merely descriptive or entertaining, nor was it didactic, self-righteous, or pedantic. Following patterns developed internationally by Victor Hugo, Charles Dickens, and Leo Tolstoy, Howells advocated for a publicly accessible fiction that modeled Christian morals and social reform. Paul Abein describes this mode of writing as imbued with "a practical sense of Christian ethics, religious tolerance, and a drive to peaceful social activism" (15). Maher's own cosmopolitan exposure to these global practices in nineteenth-century literary writing—such as his exposure to Hugo while a student in Paris—can here be seen.

In Mill Creek, Mark asserts these values by embodying them, not through confrontation or sermonizing, but simply by integrating them into his daily life and work. For example, as he travels upstream on Druker's boat, Mark politely asks about the availability of Sunday services in the area. Without proselytizing, Mark manages to compel the previously wayward riverman to recall his father's faith. Druker soon realizes the error of his ways and becomes a regular congregant at Rev. Freenberry's services. Later, when Mark realizes the extent of the company's corruption, to the point of Fieldbell's attempts to steal Mark's alkalis, he is never accusatory or angry. He simply goes about his business carefully and trusts that, in the end, righteousness will prevail. Finally, and most important, Mark's abolitionism is rooted in his Christian values. To him and the Cincinnati Republicans, slavery undermined and threatened the larger premise that the nation should become a model of Christian virtue.

Yet Mark utters these beliefs only when pressed. He cannot free or change the legal status of the enslaved miners and workers, but he can make their lives better in a number of ways. First, by improving conditions in the mines, the works, and in their dwellings—making each more efficient and thereby profitable—he prioritizes their safety and health. Second, by insisting on cash payments that allow them to pay in cash for their own necessities, he frees them to make their own choices, or even to save toward purchasing their freedom. Third, he hires a "colored Baptist minister" to offer them services where they might be most comfortable, an act that presupposes his insistence that mining and distillation operations cease on Sundays. Finally, when the opportunity comes, he not only advocates for their freedom but also takes concrete steps through the organization of a Union militia. Mostly, though, he simply treats Black Americans as humans, respecting their needs and conceding his own reliance on them, a rarity in white-authored late nineteenth-century fiction.

The unvarnished imperative of realist narrative techniques enabled authors such as Maher to present such acts with no obligation to plumb the internal motivations of the characters, by instead valuing the accurate depiction of actions and events, often at the cost of more evocative storytelling conventions. Like Mary Noailles Murfree in her collection *In the Tennessee Mountains* (1884), Maher disregarded older, negative regional conventions to focus instead on Appalachia's many postbellum transitions, rejecting stereotypes left over from the tradition of the humor of the Old Southwest.[8] In those texts, cosmopolitan narrators mocked and condescended to primitive or savage mountain folk, and the authors relied on crude debasement to imagine the absurd otherness of the Appalachians, establishing between the author and reader a bond of urban sophistication that excluded mountain folk. By contrast, Maher's Appalachian characters are never ridiculed, nor does Maher use impenetrable mountain dialect to accentuate and mock their lack of education. Local colorists like Davis, Murfree, and, in this volume, Maher seek what makes the region's insiders and outsiders more alike rather than more different.

On the whole, despite his education and worldliness, Mark treats the denizens of the Kanawha valley—white and enslaved—with respect and fairness. In the text, Druker and Johnson, though uneducated, are virtuous and honest professionals, and so merit his respect. In addition, neither would be considered racist in the moment's escalating Jim Crow–era vitriol, but rather they share Mark's relative liberal mindedness. For instance, Druker staffs his ferry with whites because slaves are unmotivated, but not because of any intrinsic inferiority: "I don't blame the poor creatures however as much as I do their masters for raising them as they do from the day they are born to be as lazy as they are most of themselves." Initially, Mark shares this view; later intimate, familial contact with slaves brings greater respect for their contributions: "taken as a whole [the leased slaves] seemed to be an intelligent-looking set of men, and, fairly treated and managed, they could be made to yield satisfactory labor as miners and wheelers, for these had all be chosen by Johnson, and even as chargers." In fact, treated humanely by Mark, they strip off the masks they had donned to survive the indignity of slavery and reveal the intelligent and diligent personalities they had previously concealed. Mark's own reliance on John as his spy transcends the emergent open-mindedness of the mountain whites to reflect a more fully expressed postslavery model of racial cohabitation and cooperation.

Throughout, Maher describes Mark as working to create a genuine community in Mill Creek: one in which neighbors of different races, backgrounds, classes, and genders might live together in peace, a subtext designed to resonate with his readers in 1887 as much as it reflected the community of Mill Creek in 1860. However, the mining communities of the Kanawha valley were, historically, a fortunate and apt setting for Maher's progressive local color script.

Miners and Slaves

In *Black Coal Miners in America: Race, Class, and Community Conflict* (1987), Ronald L. Lewis begins with a section on leased slaves working in the coal mines of Appalachia, focusing on the Kanawha valley in particular. Before 1850, as many as 3,140 slaves mined the region's coal to fire the salt furnaces, an industry that was soon in decline. After 1850, most leased workers were switched to cannel coal mining and processing as forty-six coal oil companies opened in the region between 1847 and 1861, according to Lewis (8). In the novel, Reynolds's abusive treatment of the leased slaves prior to Mark's arrival seems typical of the region's practices: Lewis even describes one company that penned up their leased slaves at night to keep them from escaping to Ohio (11). In the novel, Reynolds never goes that far. Yet under his domain, to replace their clothes and other basic necessities, the slaves were given simple notes to Charleston merchants, and had no choice as to their clothes or shoes, resulting in the rags Mark found them in, thanks to Fieldbell's skimming of the available funds. Unlike leased slaves elsewhere, they were denied the opportunity to save their own cash, either for satisfying other needs or for saving toward buying themselves. Small wonder the enslaved woman, Maria, brought to cook for the miners had returned to eastern Virginia. But upon Mark's arrival and the implementation of his reforms, she returns.

Unlike their treatment by the company, the slaves and former slaves in the novel, like John, face little stigmatization from Appalachian whites. Lewis explains the source of this ambivalence: in the Kanawha, "slavery was primarily an economic matter of 'cheap' labor. Since the institution of slavery was not the foundation of western Virginia society, there was no overpowering need to demonstrate ideological allegiance to the institution" (12) by the white settlers. Local whites did not resent their presence as the labor market had no surplus, and besides, mining was dangerous and unpleasant work. In the novel, the whites easily admit themselves abolitionists and Republicans simply because it reflects their Christian values.

Once they have settled in Mill Creek, Mark and Sarah employ Alice, John's sister and Maria's daughter, as a domestic servant, and pay her, although she is technically a leased slave. The intimacy of the two families is apparent from the start, as the slaves come to trust the Marks with knowledge of John's secretive travels. The Mark children teach Alice and Pole, a slave employed in the works, to read, and allow Maria and Alice to share in conversations between Mark, Sarah, and the Johnsons. John's role in arming the militia demonstrates trust and confidence in both directions. Very few people in western Virginia actually owned slaves, only 1 percent, according to Mark A. Snell (17). More like that of northerners across the Ohio River than the southerners across the mountains, western Virginia's economy did not depend on slave labor, and white identity was not defined simply in opposition to slaves, as W. J. Cash long ago suggested was central to poor white identity elsewhere.[9]

On the whole then, with regard to race, economics, industry, and other reasons, western Virginia had diverged from the Richmond-based east long before 1861 in both documented history and in the novel. Small wonder that when the crisis arose, western Virginia voted to become West Virginia.

Secession and War

On studying the sources of Virginian intrastate divergence in the 1850s, Richard H. Owens notes that "citizens in the north and northwest sections of the state . . . complained that they were treated as the vassals of the eastern politicians. They believed their taxes were levied and increased for the benefit of the easterners. Internal improvements, such as canals and railroads to connect western areas to eastern markets or as links to the Ohio valley and western states were always a divisive issue. . . . The improvements remained more planned, promised, and largely unbuilt than realized or actual" (13). For example, despite eastern promises, the Kanawha River had not been made more navigable through dredging and banking, as had been the rivers used by mining companies in northern states like Pennsylvania (9). Like the Whiskey Rebels of the 1790s farther up the

Ohio valley, western Virginians queried their loyalty to the eastern authority that expected their obedience, and their taxes, even before the war.

In their comprehensive *West Virginia: A History*, Otis Rice and Steven Brown lay out in great detail the process by which western Virginia seceded from Virginia itself, as does Maher in the novel. In chapters seven and eight, Mark repeatedly reads to his family, the workers, and the leased slaves from Republican papers from Wheeling and Cincinnati delivered by Druker, the white boatman, and by John and Isaac, the self-freed ex-slaves.[10] The western counties held a popular referendum, as Maher describes, on May 23, 1861, a little more than a month after Virginia had voted to secede from the Union. The vote to secede from Virginia was very close, though not in the Kanawha valley where it cleared by a large majority. On May 27, Union general George McClellan crossed into West Virginia, as did other forces, such as the Ohio regiment that eventually rescued Maher. The new state did not immediately ban slavery and so existed as a "border state" during the war, unaffected by the Emancipation Proclamation of 1863. Only after the passage of the Thirteenth Amendment in 1865 was what little slavery that remained banned from West Virginia.

True to the postbellum spirit of reconciliation and his own Christian ethic of forgiveness, Maher writing in 1887 refuses to wave "the bloody shirt" to demand the punishment of the South for the war or for his own brief captivity. Instead, Maher depicts the Southern soldiers sympathetically, such as the one convinced by Mrs. Johnson to return home. Maher only depicts the leaders as a problem: General Henry Wise and Major Crone.[11] Former Virginia governor Wise had been sent west on a recruiting mission. Druker articulates the local response: "The old fool. . . . He don't know us yet with all his knowing. The Kanawha is not Richmond. Let him try and send his Grey chaps or green horns up the Elk, and if they don't learn to throw their guns and run like hounds before they have made four mules, my name is not Druker!" When he realizes this, and anticipating the arrival of Union forces, Wise retreats. He had miscalculated the west's difference from the east, demonstrating the Tidewater region's wholesale ignorance of the western part of the state. As such, he is a nuisance, though never immoral or violent.[12]

On the other hand, Crone, who invades Mill Creek with the expressed intent to capture Mark, and eager to incur violence to do so, is a dangerous drunk. When Crone's men arrive to capture him, Mark realizes that Crone's intoxicated notions of glory would lead only to the suffering of his neighbors and the possible destruction of the mine and the works, and he surrenders. Reflecting Maher's views on temperance—again, stated but never sermonized—Crone's poor and dangerous decisions are based on a moral flaw, not his Confederate loyalties. In the postwar moment when he was writing, Maher again imagines a premise for national healing rather than further division, but only under the terms of moral rectitude and repair. While Maher depicts slavery as a moral sin repeatedly throughout the text, Mark consistently regrets the seeming inevitability of war as he discusses it with his children and others. His and Sarah's activities in the politics of secession, however, would attract public attention decades later.

Sarah's American Flag

While most of the novel fictionalizes Maher's own experiences, near the end, an event transpires which merits particular attention as it diverges from later accounts in important and telling ways. As noted above, there were no American flags in the Charleston area to rally support for seceding from Virginia. In the novel, at Sarah's demand that she sew a functional Stars and Stripes, Mark enlists John to bring from Wheeling the necessary silk bunting. With the help of Mrs. Johnson, Sarah sews the flag, and Mark and Sarah take it to Charleston, where he speaks on a stump near the polling station, and helps the local vote go heavily to separate from the "east," create West Virginia, and stay in the Union.

However, in a series of other, much later versions of the story published around the turn of the century by newspaper journalists, the contributions of both John and Mrs. Johnson are omitted. The longest of these (appendix A), based on an interview with Theophile and Sarah in 1901, was published in the *Evanston (IL) Press*. An anecdotal history of the flag and Sarah's obituary (appendices B and C) likewise recount versions that exclude Black and working-class roles to what is presented as an act of unqualified patriotism and heroism emanating exclusively from the white, educated class. In turn-of-the-century America's print culture—which championed Jim Crow legislation, the violent suppression of labor organization, and lingering Victorian sexism—the journalists who reproduced the story of Sarah and the flag simply excised both John and Mrs. Johnson from the narrative.[13]

The *Evanston Press* story characterizes Theophile and Sarah as simply charming old folks, and given the newspaper industry's enthrallment to corporate interests at the time, such erasures seem likely: it is a more marketable story if a charming and comfortable old white woman in a picturesque Wilmette cottage is the lone heroine of the story but only if she acts as an adjunct to her husband's more direct and public patriotism. The novel's version of the story speaks to the inclusive nature of Maher's authorial intentions and so reflects a more inclusive model. By contrast, the novel downplays Theophile's stump speech in support of the cause of West Virginia to focus more on the larger political contributors to the cause more generally. Regardless of which version of the story is factual, to Maher's credit, *Cannel Coal Oil Days* memorializes the flag-making event as a collaborative effort that involved Blacks, working-class whites, and well-spoken women.

Cannel Coal Oil Days is a lost though somewhat flawed treasure, now found. Starting in the 1950s, the work of recovering so many lost writers and their books by literary historians has radically altered and broadened our understanding of America's culture and history. These efforts have brought a vast diversity of voices and experiences into the tapestry of the nation's self-exploration and self-expression. More artifact than art, this novel brings to life a long-lost moment of transition and transformation. Appalachia has long suffered in the national story as a backwater, a place well behind the rest of the nation with regard to modernization and progress. *Cannel Coal Oil Days* offers an alternative Appalachia, one where the mountaineers fight for freedom, and features a coal company engineer and manager determined not to exploit the workers and the land for the sake of immediate profit. Although a cosmopolitan outsider who had traveled the world, Maher found up the hollow a place of opportunity and innovation, and not just a testing site for innovative mining technology, but also a proving ground for social progress and moral reformation. Given all the experiences he could have addressed—from Paris to California—Maher found the story most worth telling among his friends and colleagues in Mill Creek, West Virginia.

Notes

1. On the impact of Creole refugees from Haiti (St. Domingue) in nineteenth-century America, see Goudie. For more family history, see George Maher's Wikipedia page: https://en.wikipedia.org/wiki/George_W._Maher.
2. Throughout Gesner has proved invaluable for a source on nineteenth-century coal oil technology and history. Other useful sources are Erni and Graham.
3. Throughout, Lewis has been consulted for history of enslaved miners in the region. On slaves and secession, see Rice and Brown, 141–49.
4. See Hatch and Abzug for overviews of progressivism's roots in the Second Great Awakening.
5. In 1853, Harriet Beecher Stowe employed "white trash" to refer to impoverished, uneducated, and often unemployed Southern whites.
6. As well as Erni, Gesner offers the best explanation of the techniques and equipment referenced by Maher (12–15). Moreover, Gesner's brief volume describes the processes and offers illustrations of the equipment. It was intended as a guide for

"chemists" who had embarked in the field. I have consulted both Erni and Gesner throughout on the history of coal oil production.

7. See Venable's *The Beginnings of Literary Culture in the Ohio Valley* and Watts's *An American Colony* for a more complete discussion of the culture of antebellum Cincinnati.
8. See Booth on the realist tradition in American coal mining novels and the genre's tensions between locals and outsiders. See Justus on images of Appalachian and Southern regions as backwards, violent, and ignorant in the Philadelphia-based genre of humor of the Old Southwest.
9. See Cash's *The Mind of the South* (29–58). See Dorsey on West Virginia's and Appalachia's struggles with overcoming these issues.
10. On the formation of the state of West Virginia, see Owens (21–28) and Snell (17–28). Rice and Brown also provide extensive discussion (140–70).
11. Biographical information on Wise can be found in Hambleton and on his Wikipedia page: https://en.wikipedia.org/wiki/Henry_A._Wise. No information is available on Crone.
12. Rice and Brown detail this process at length in their section of the Kanawha region (122–34).
13. See Smith-Rosenberg, 167–81, for a discussion of the exclusion of women in turn-of-the-century public discourse.

Editorial Note

The existing manuscript was intended as only the novel's first two-thirds. The rest has either been lost or was never completed. No other trace of it remains in Maher's papers or in other family or archival records. It was scanned from the handwritten manuscript using Transkribus at Michigan State University and edited for modern usage, orthography, and punctuation.

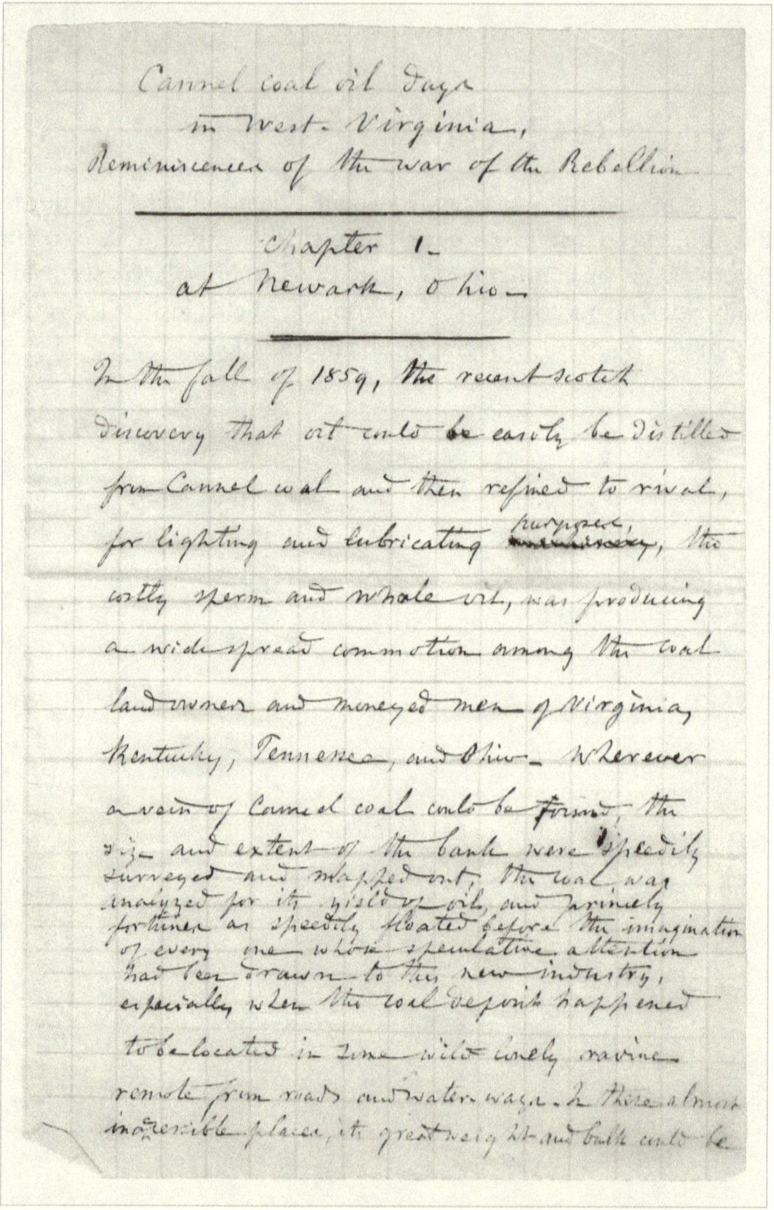

Cannel coal oil days
in West-Virginia,
Reminiscences of the war of the Rebellion

Chapter 1.
at Newark, Ohio.

In the fall of 1859, the recent scotch discovery that oil could easily be distilled from Cannel coal and then refined to rival, for lighting and lubricating purposes, the costly sperm and whale oil, was producing a wide spread commotion among the coal land owners and moneyed men of Virginia, Kentucky, Tennessee, and Ohio. Wherever a vein of Cannel coal could be found, the size and extent of the bank were speedily surveyed and mapped out, the coal was analyzed for its yield of oil, and princely fortunes as speedily floated before the imagination of every one whose speculative attention had been drawn to this new industry, especially when the coal deposits happened to be located in some wild lonely ravine remote from roads and water-ways. In these almost inaccessible places, its great weight and bulk could be

Fig. 5. First page of original manuscript (editor's family collection)

CANNEL COAL OIL DAYS

Chapter One

At Newark, Ohio

In fall of 1859, the recent discovery that oil could be easily distilled from cannel coal and then refined to a fuel purposed for lighting and lubricating, though more costly than sperm and whale oil, was producing a wide spread commotion among the coal land owners and moneyed men of Virginia, Kentucky, Tennessee, and Ohio. Wherever a vein of cannel coal could be found, the size and extent of the banks were speedily surveyed and mapped out. The coal was analyzed for its yield of oil, and princely fortunes as speedily floated before the imagination of every one whose speculative attention had been drawn to this new industry, especially when the coal deposit happened to be located in some wild place remote from roads and water-ways in almost inaccessible places. Its great weight and bulk could be reduced from a bushel to a gallon and be made thus able to reach distant but ready markets in all directions at a low cost for transportation.

Vast sums of money had been freely spent for the construction of large works for this distillation and refining at the mines that had been opened, and also for roads to connect them with the nearest navigable streams or with the nearest railroad. Chemists, druggists, physicians, and even schoolteachers were sought after, on account of their presumed acquaintance with chemistry, to advise the most advantageous modes of reducing cannel coal oil. The usual feature in a mad race for riches in sight was not wanting in this, namely the employment at high salaries of unworthy and unprincipled characters proclaiming themselves as discoverers and masters of the art of a superior process for distilling and refining beautifully yellow colored and pure cannel coal oil.

Petroleum was not known then to exist in impure quantities in the bowels of the earth, and although faint rumors were already finding their way to the west from Pennsylvania that hardy adventurers were there at work hunting and boring for it wherever they spied it trickling down from the rocks or springing out of the ground. Even more adventurers were being supported by small samples sent in vials to show how much clearer it was in its crude state than the tarry blackish liquid distilled from this coal. Yet no fears were allowed to be entertained of a serious competition from that quarter.

An operative chemist landed then from the car in Newark, Ohio. His wife and three children were with him. They had come from Chicago where a fire had just destroyed a large establishment which he had constructed but a few months before to manufacture the chemicals needed by a cannel coal oil manufacturer for separating it from the impurities usually removed by the retorts. This fire had left him almost penniless but the practical experience he had gathered in the preparation of these reactives made him bold to feel that he could make use of it where it was evidently needed, and his brave wife shared also his hopes. Packing their furniture, they had forwarded it by mail to Newark and preceded it in a passenger train. They had passed through a similar trial by fire before also in Chicago, and previously in California. There, an establishment of the same kind which he had erected for the separation of gold from silver had been wrested from them; so that, with their children, they had been compelled to return to their home in Philadelphia.

At their arrival, to their utter amazement, they had found their property there suddenly locked up in a prior litigation in congruency of the inconsiderate act of the wife's brother. All their trials however had been accepted by them with thoughtful reverence. Their trust in the promises of a covenant-keeping God had not been shaken. It had been strengthened, as the roots of the oak, through storms and rains, go steadily every downward deeper and deeper every day and night into the ground to send up from it into every branch and twig the nourishment and force they need to bow before the tempest without breaking, and losing their hold upon the trunk. Tested, they had been confirmed in their assurance that His love is that of perfect wisdom and that He knows best what is best for all whom He loves.

Within a short distance from Newark were located several cannel coal banks, which were being extensively mined to supply the retort benches and line of huge stills connected with them. Mark was his name, and he was personally a stranger to these Newark coal oil manufacturers, but he had, from Chicago, sold and shipped to some of them large quantities of the concentrated acid which they needed for their refineries. The other reactive, the alkali, they prepared themselves in their works, but it was of an inferior strength. They were thus unable to obtain the pure oil which he showed them they could produce by using a superior strength reactive unknown to them and prepared by him in a small way. His practical acquaintance with the intricacies of this manufacture was thus speedily appreciated. His proposal was accepted by two well-known merchants of the town to contract and operate the first small works to supply this strong alkali. Some of the local refineries used it to enable them to send out a finer oil, much to the ill-disguised confusion of certain professors. Mark, having at one time been employed as an assistant on a state geological survey, had succeeded in selling his improved process at an extravagant price to several of these oil companies, demonstrating his mastery of the art of distilling and refining cannel coal oil.

Mr. Mark, in making known so freely a newly proven technique which had cost him much time, study, and technical testing while in Chicago, wished to facilitate the labor of these refineries, and render it more remunerative. He wanted also to create in Newark a larger plant to offset the costly manufacture of acids, which were now being obtained from Cincinnati at a much higher price since the destruction of his own works. His plain unvarnished method and the superior results obtained through his counsel had won him the friendship of all these oil manufacturers. The attention and desire aroused at first for this additional construction were, however, during the winter held in abeyance by an increase of production from other mines, and also by continued reports from Pennsylvania that petroleum was no longer a myth but a reality. Wells over flowing with this clear and easily refined oil were being bored with increasing rapidity and even worked up in refineries erected near them.

From the Kentucky owners of a rich mine in Virginia, ten miles back of the Kanawha and the Kanawha Court House, a letter came to one of these Newark friends inquiring for a manager whose qualifications he could recommend. "Mr. Mark," said he one day as they were discussing the situation, "you have rendered to us all good service here this last winter, but you have hardly benefitted yourself, if I compare your generous course with the grasping methods of that doctor yonder who has fleeced so terribly every one of us these last eighteen months experimenting to learn all he knows at our expense. Our mines here are small and poor compared with those in Virginia. The oil men who exploit them there don't know what you know, what you have taught us. That is the field for you, one far more promising than this. How would you like to try it and take hold of the large works in the Kanawha owned by some old friends of mine? Here is their letter inquiring for such a man as I know you to be, for managing their property. The writer of it lives at Livingston, Col. Truesdell, the secretary of the company, but he is really the main owner, being the heaviest stockholder. He bought the land in the first place, and he organized the company two years ago from among Kentucky friends of his."

"Who is the President, Mr. Shortmoore?"

"The President is a Mr. Fieldbell. He stays at Kanawha Court House, runs the refinery there, and thence manages the mine and retort works on Mill Creek ten miles back of the refinery with which it is connected by a boat on the Elk River, excepting two miles of a wagon road through hills and woods. The company are not satisfied with his management: they complain of his expenses and of his not turning out as much oil as they think he could, nor is it as good as it ought to be. The little share he owned was presented him as well as the Presidency, and a high salary to encourage him to make good his word to them when he took the job, recommended as he was by Col. Fynch of Ohio, a large-stock holder himself and a relative of his wife. Mr. Fieldbell was the principal of a high school and a *savant*, something like our Professor here, and has learned doubtless like him all he knows now of coal oil at the expense of his employer."

"Who is the treasurer?"

"The treasurer is a nice old Christian gentleman, Col. Albert of Frankfort, Kentucky, well known and highly respected as the state printer since the days of Henry Clay. From what I have seen of you, I feel sure that he would capture you by his very methods and straight ways, and you both would make a good team. They have an office in Cincinnati."

"What about my present work here?"

"Well, your investment is not heavy, and, in our present state of suspense here, from some late conversations with the two gentlemen who have joined you and helped you to put up and run your small alkali works as an experiment, I think they would be willing to sell them to me at cost, if you are willing, after having satisfied yourself that the change would be one you would like to accept."

"Well I will speak to them on the subject and to my wife also, I thank you. Thank you, Mr. Shortmoor, for your kind consideration."

"Mark, one good turn deserves another. Let me know your decision as soon as you can, that I may answer this letter and inform my friends when they can look for you."

One of Mark's two associates had just returned from an exploring trip to the wild localities in western Pennsylvania where petroleum was being sought for eagerly by hundreds of oil hunters well supplied with money to pay for their derricks and borers. His conclusion from all he had seen and heard was that the small expensive mines of Newark, with all the legal disputes which encumbered them, would render their competition with natural petroleum perhaps a hopeless one in the end. It would therefore be a dubious move for himself and his partner to embark any more of their hard-earned money in an enlargement of production for which there might eventually be no longer any paying market. His partner agreeing with him, Mark stated to them the proposal of Mr. Shortmoor, and a friendly understanding was reached that, before his offer should be accepted, Mark would go down to Cincinnati and Kanawha to explore both grounds and to confer with the Kentuckians. The good wife also assented. She was no longer among strangers: their cottage home in the outskirt of the town was commodiously arranged, pleasantly surrounded by fruit trees in full bearing, and flanked by a garden, which she could get cheaply worked by one of the men of the alkali works nearby. The rent was very moderate, and every day or two the mail would bring her a letter from her husband telling her progress.

He reported all this to her. The day after, kissing his wife, his little son then five years old, his sister fifteen months younger, and their little younger brother but a year old, he stepped on the car bound for Cincinnati.

Chapter Two

From Newark to the Kanawha

The prospect before him was dim. His hopes of success in Newark had not been realized. In going to Virginia, he would meet strangers again—the leading spirit among them whom he was called to supplant, vested though he was with the authority and power of the President of the company. He hated strife and was going deliberately to face it. He knew well however how and where he could obtain the sound judgment he lamented he had lacked before when subjected to similar ordeals, but which he felt he so much needed now to give strength and unquestioned command to a conscious, well-balanced honesty of intentions and way. "It is not all to mean well. We must act well also," he would say to himself. From the hotel in the morning, he went to the office of the Kanawha Company. Introducing himself to Col. Albert, the treasurer, they soon entered into a long conversation in the midst of which Col. Truesdell, the secretary, arrived from his house in Lexington, across the river. Truesdell was a distinguished lawyer and politician also, and he eyed sharply the newcomer, sizing his mental measure and weight as well as he could from his past experience with men and seemed satisfied, after joining in the conversation, that his Newark friend had not failed him.

"We have a fine cannel coal property on Mill Creek, Mr. Mark," said he. "Two thousand acres well-timbered with oak, walnut, chestnut, poplar, and beech, ten miles from the Kanawha and two miles from the Elk which empties into the Kanawha just below the courthouse. There we own also ten acres on the Elk upon which, on a bluff, we have built our large refinery, and to it our boat takes down twice a week our crude oil in barrels which are hauled in wagons to the Elk from the works on Mill Creek. Our intention, as soon as we see the way clear, is to lay down a tramway of two miles to connect our retorts and mine with the Elk, and thus save our present heavy cost of hauling with two teams of horses. We will of course pay all the expenses of your trip to both works, so you may look at them and decide carefully and from your own judgment, if the methods by which they are now being managed and upon how you could improve them to reduce expenses, and even, if possible, to increase production with present pictures without calling on us for more cost of which we have had enough already, excepting however this tramway which we must have. All this you can report to us when you come back. The boat leaves our wharf for the Kanawha tomorrow morning, Wednesday, and will reach the Court House Friday morning.

"With a letter of introduction which Col. Albert will give you for Mr. Fieldbell, you will call on him at the refinery, make his acquaintance, and proceed with your investigation. Mr. Shortmoor has told you how we stand in our relations with him?"

"Yes, sir, and I have to act with no little caution in my intercourse with him."

"Well, he gets a high salary, and ought to do better. This we know by comparing the capacity of our works with the capacity of other works on the Kanawha and their production. We look to you to give us your proofs. In trusting this matter in your hands, we rely much upon Mr. Shortmoor's judgment and his acquaintance with you. The Kentucky friends who have invested with us in this enterprise have done so upon our representations to them of the value of this stock, and you understand our responsibility to them. They all have other business to attend to and look to us for good management and their shares of the net profits. How long a time do you think you will need to form a correct judgment and report to us?"

"Two or three weeks, I presume will be sufficient."

"Well, this fifty dollars will pay your way till you return, and we hope that this trip will prove advantageous both to you and to us."

On his return to the hotel, Mark wrote to his wife an account of his interview and his hope of a favorable outcome from his conversation with the two Kentuckian gentlemen. He then went down to the Kanawha boat, engaged his passage, had his trunk taken down to his state room in the afternoon, slept on board that night, and, in the morning, the *Ellen Fray*, such was the name of the boat, was plowing her way up the Ohio River. It was in early April, a soft spring breeze was wafting from the west, the chirping of birds could be heard from both shores, the green fields and orchards which lined both shores and their appearance suggested to the passengers many comparisons between the free state of Ohio on one side and the slave state of Kentucky on the other. The many landings for passengers and goods afforded opportunities also to observe the many points of difference in the character and habits prevalent in the two states.

The loud claim of the South that slavery should be recognized as lawful property by the North in the territories as well as in the states had come to the front as a national question, and the right of this claim was being as stoutly denied at the North. So emphatic also was the protest, that both shores seemed painfully, and also anxiously, alive as to the immediate and ultimate results of the impending clash. Mark's mind was deeply moved as he brooded upon the question: how long will this enchanting scene of peace, of plenty of quiet, happy homes remain unbroken by the terrible dim and devastations of civil war? How long before these green fields are turned into a barren waste dotted at this bend and that bend yonder with frowning batteries and forts menacing each other from either shore, stopping all boats like this and their trade, and hushing with the stillness of death all the activities which now charm the eye as we pass up from farm to farm? The greater portion of the passengers were Virginians from the salt licks and furnaces on the Kanawha or traders living on that stream whose early limited training and whose family ties naturally bound them to the public sentiment of their state. On the other side, lifelong commercial associations with Cincinnati and other towns on the Ohio had enlarged their mental visions to see also that the authority of a constitutional inseparable union must be paramount to a mere assertion of States' Rights in direct conflict with that authority for the universal recognition of a barbaric right of ownership in human flesh.

The boat reached the courthouse on Friday morning, and Mark, following the Black porter who had shouldered his trunk, went up the steep landing to the hotel. He had hardly written his name on the register, when a middle-aged man, who stood by and could read it on the book, spoke to him: "You are Mr. Mark, sir, are you not?"

"I am, sir—"

"My name is Fieldbell, and I see, from this letter by mail that you have a note of introduction for me from Col. Albert of Cincinnati?"

"I have, sir," said Mark, taking it out of his coat pocket and handing it to him. "I am glad to meet you here so readily."

Fieldbell read the few lines, and entered at once into conversation with Mark, inquiring how long he had been engaged as a manufacturing chemist, in what branch of the chemical industry, and what particular were the processes Mark had used to extract illuminating oil even from the tar refuse at the gas works at Chicago, a feat which he thought would hardly pay for the cost.

"It does pay," said Mark. "At the present market rates for coal oil, and in connection with the other products of this distillation, for they have all a specific value." He named them in rotation. This evidence of practical familiarity with an industry which Fieldbell professed to have mastered was not tasteful to him, and Mark, on his guard already from his knowledge of the man, saw that he could hardly hide his displeasure.

"Our friends in Cincinnati," said Fieldbell, "are not pleased with my production. They want more oil from me, but I cannot refine more than is sent down to me from the mine, and the men there tell me that they work all the coal they can with their one hundred retorts. Will you come down with me to the refinery after dinner?"

Mark assented and took his leave for a while to go up to his room and to his trunk which had been carried up. He opened it to satisfy himself that the two small bottles, in which he had brought his reactives to test the crude oil, were not broken. He found them safe and whole, glanced at the grate filled with coal and kindling ready to be fired, and, hearing just then the sound of the dinner ball, went down to join Fieldbell who was at the foot of the stairs looking for him. Conversing together at the dinner table, Mark inquired how many companies on the Kanawha were engaged in mining cannel coal to distill and refine the oil?

"There are twelve," answered Fieldbell, "all, like our own, working high rich veins, not low veins and slaty like yours in Ohio, and most of them have more retorts than we have, using of course at their refineries more acid and alkali than we do. This would be a fine field for you, to put up such works as you wanted to construct at Newark." And he looked sharply at Mark, who remembered instantly the words of Shortmoor concerning the Newark professor. But he had not come for this at the expense of the Kentucky company. They had sought and secured his services for their own guidance in the conduct of their large investment. Returning the sharp, inquiring look of Fieldbell, he replied, "A large capital would be required to build works equal to this supply, and I am a stranger here."

"Oh! You could soon obtain it right here."

"I am on another errand, Mr. Fieldbell."

"Yes, yes, I merely suggested this for further consideration."

They had thus both caught a glimpse of each other's mind, and how it was Fieldbell's turn to be also on his guard. After dinner, they walked down about one mile along the main street which bordered the Kanawha, until they reached a turn into another street which bordered also Elk as it flows into the Kanawha. Then they went on a half mile further, to a large open field overlooking the Elk. Between these fields, at the foot of heavily timbered hills, and the Kanawha, the old-looking narrow town stretched itself upon a length of two or three miles. The refinery stood inside of this border, a huge one-story frame structure, looming up about fifty feet over the water of the Elk. A flat boat upon its edge was moored to the heavy timber of a slanting wheelway on which barrels of crude oil were then being hauled up on large flat cars by a cable fastened above to a capstan steadily moved by steam power.

Mark thought at once that this expensive and awkward mode of transportation would be much simplified by the work of a steam pump here, and the use of a deck-covered boat into which, at the upper landing on the Elk, the crude oil could be run down through a large hose or pipe from an iron tank hauled on wheels from the retorts at the mine. This suggestion however he reserved for his report. His next subject for observation was the large size of the iron stiller, agitators, and tanker, one half of which were idle, unused, but evidently intended, in the original plant, as it had been handed to Fieldbell, either to refine all the crude oil from the mine or to increase production from an enlargement of the works there. He would not well elucidate this point from Fieldbell who passed over these idle fixtures with the slight remark that they were used occasionally to relieve a press of work. Mark's mind was not slow to realize the fact that the work at the mine was sadly neglected or in incompetent hands. If the distillation and refining processes were all what they should be to produce a fair clear oil, then as he saw it, the oil to run into new barrels could be shipped to Cincinnati on the same *Ellen Fray* which had brought him.

As his process constituted the main ground on which Fieldbell had from the start staked his claim on a very high salary, Mark remembered the tactics of which the professor had warned him at Newark, but said little on the subject. He filled however a quart bottle with the crude oil which had just arrived in order to test it on his return to his room at the hotel.

"Have you the still, the instruments, and the reactives for this test, Mr. Mark?" inquired Fieldbell curiously.

"I have," Mr. Fieldbell.

"At your room? In your trunk?"

"Yes, sir."

"But what will you do with the refuse which you will want to throw away, and the stench from it in such close quarters?"

"Oh! Practice makes perfect, you know. It will not be the first time that I have sought this kind of knowledge, under difficulty without causing any trouble to anyone."

Fieldbell made no reply. He went on with his work, and Mark made his way back to the hotel. Their conversation had been partly overheard by the men employed in the building and by the boat men also. At Mark's request these last were ordered to remain until the next morning, when he would have his trunk brought downstream, and he would go up the eight miles with them to the landing on the Elk, they poling and pushing the boat loaded with empty barrels. He could thus, in his report, explain the labor which could be dispensed with by the adoption of an obviously more rational and less costly, less laborious, and speedier mode of transportation.

On his return to the hotel, he debated in his mind whether, during the two hours yet before supper and also after supper until bed time, he could not, in his room, complete the test of which he had spoken to Fieldbell. He had just concluded to commence it when a knock at the door stopped his cogitation, and, closing his trunk, he unlocked the door and stood facing the landlord in person whose anxious eyes rested at once upon the black sample in the bottle which Mark had laid upon the mantel piece.

"You are a chemist, sir, I understand? You are not going, I hope, to set fire to my house with any experiments here?" Mark laughed, understanding the frightened movement of the scared men at the refinery, admiring also the quickness of motion and how well he had devised to impart their fright to the worthy old Virginian by some penciled warning dispatched by some Black boy on a run through one of the back streets.

"Oh, no! Sir," he replied still holding the door. "You need not be alarmed. Ten miles away from here tomorrow or the day after, I will have a better opportunity to experiment on this black specimen of crude coal oil."

"Are you going away tomorrow?"

"Yes, I am going up the Elk on an oil boat and will get you to have my trunk carried to the refinery. I will go with the porter." He had thought at first that, in his first letter to Cincinnati which he wanted to write in the evening, he would have been able to add a record of his first test of the quality of the crude oil as sent from the mine. But he would explain his want of time for the trial, after reciting to Col. Albert his visit to the refinery, what he had seen there, and his object in pushing up at once to the crude oil works by the boat. When the supper bell rang he went down, and, as he passed into the dining room, he could see Fieldbell standing at the door of the hotel and talking a part with the landlord in an earnest but low tone of conversation, looking shyly also toward his guests in his main hall, who, in response to the call of the bell, were moving in lines to their seats around the table.

"Have you started your test, Mr. Mark?" said he as he sat next to him.

"No, sir," answered Mark smiling. "There has been a scare here. The landlord somehow divined my intention and came up just in time to put in his view with his eyes looking straight at our black bottle which he had spied upon the mantel piece. I of course made no objection and bowed him off."

"Mill Creek will be a more favorable place than this house for chemical investigations. Reynolds, our engineer there, who runs the works, has only two rooms in the log house which he occupies with his wife since I have been here. How long do you intend to stay with him?"

"I do not know yet. Have you written the few lines I want for him?"

"Yes, here is my note to introduce you to him. I instruct him to afford you all the facilities which Col. Albert wrote to me that you should have for your investigation. You will find him a slippery fellow rather, fond of his own ease, though an experienced engineer. The Blacks, all slaves, whom we hire by the year to mine the coal and run the retorts, know that he is not overly fond of work. He does not watch them, and they treat him as much as they please in the quantity of coal what they dig and wheel out, in the measurement of the coal wheel they charge the retorts, and in the firing of the retorts between charges. It is high time that he should go back to Cincinnati where Col. Truesdell found him at the foundry works where all our fixtures were bought. You will soon have his weight in your mind. His wife however is a very clever woman, a better housekeeper than he is a manager; and I am pretty sure that he is feathering his nest not only with his big salary, but with all he can make off the extras which he says he has to pay every month to the Black hands for over work, as he puts it."

Here was another revelation, which Fieldbell freely let out, hoping, as Mark shrewdly surmised, that he would discover flaws enough at Mill Creek to keep him permanently mending them, and leave him no time to spare for the refinery, especially if the production of crude oil could be so brought up as to call into full play, the year round all the capacity of the refinery.

"Well," said Mark, "this is bad enough. How will I get through with such a man to lean on for information, correct information?"

"I think you can master him, Mr. Mark, from the little I have seen of you." Mark bowed and smiled, but said not a word, and finished his supper silently, thinking of all the dark days before him, so suddenly, but so forcibly described by one whose authority ought, long before this, to have been used for the removal of perhaps the main cause of so much waste and so much disregard to the large vested interests. The greater the distance, he thought, between the owners and their property, the more keenly should Fieldbell have felt his responsibility for them here.

"Can you be down between seven and eight o'clock tomorrow morning, Mr. Mark? The boat will have to make an early start to reach the landing in time for the men to unload the empty barrels."

"I will be there between six and seven, Mr. Fieldbell."

"Very well, the captain of the boat will be here this evening and I will tell him to be ready to start at seven. Do you want to see him?"

"No, I have some letters to write home, and you can make the arrangement with him. What about the wagon at the landing to carry my trunk to Mill Creek?"

"The wagons may be there when you arrive, or their second trip. If not, Capt. Druker of the boat will accommodate you for the night at his house close by the landing till the morning when you will not mind walking the two miles to the mine along Mill Creek though you would have to climb it more than twenty times on logs and gully-water. It was a wild country when the mine was opened, and the works built there. The road had all to be dug along the base of the hills which border the creek to the landing on the Elk, at a heavy cost for blasting rocks and cutting big trees." They parted for the night.

A letter to his wife and another to Col. Albert occupied Mark's time until wearied nature called for sleep. To his wife he related minutely this second stay of his journey as he had detailed to her the first at Cincinnati. He told her the discoveries he had already made and his anticipation of still sadder ones which would probably result in their being hedged up with their children untutored in a wild country and among people far away from the House of God and any school, just at that tender age when impressions received are most lasting to form character and habits. It was a trying perspective for him to contemplate as a husband and a father. But he was writing to a trusting heart like his own and as Providence would lead them in their extremity, and he was determined to follow this path, especially if this Col. Albert to whom he was going to write would answer him in such a way as to win his confidence and respect.

He wrote to Col. Albert unreservedly all his impressions of his call at the refinery and all his conversations with Fieldbell from the moment they had first met to their parting after supper. He had remarked at Cincinnati the deference paid by Col. Truesdell to the judgment and the superior age of Col. Albert, while the latter seemed to evidence a sort of fatherly pride in the business enterprise and capabilities of Truesdell whom he and his wife in Frankfort had, he understood, much befriended him when, a young lawyer, he was struggling for advancement. In closing his letter, Mark wrote to Col. Albert that a few days might elapse before he could gather sufficient reliable information for a third letter from Mill Creek.

Chapter Three

Up the Elk River to the Landing

At six thirty in the morning Mark was on the boat, seated upon his trunk, looking at the men preparing for their return trip. Capt. Druker caught his eye as he was measuring with it the height of the bank upon which the refinery was built, and said: "Mr. Mark, this Elk River is very apt to rise in a few hours after a heavy rain to within a very few feet below the top of this bank, and then falls again thirty and forty feet as low as it is now. Over the Kanawha yonder just below the bridge, there's another mountain stream. It's hard work to watch our lines day and night or shorten and lengthen them when the boat is fastened to this wheelway or to the other up at the landing."

"How do you manage to pole up against the current when the river is high on your return trips?"

"Oh! We take to the rope then and use a horse on that bank to pull up close in shore."

"At this low stage are there any rocks in the way, any shallows?"

"Yes, we have then to jump out and push the boat through them, full or empty. It's hard work in winter. In summer sometimes we take in a horse to pull us through with our load of empty barrels."

"How do you dry your clothes in winter when you jump out and in again?"

"We keep a small pile of sand on that rudder bench and build upon it a small wood fire with pine knots."

"What if it sparks when a high wind blows upon the oil barrels?"

"Oh, we watch for that. We open and stand up this sheet from around our fire."

This odd way is not very safe, thought Mark again. A small, rough cabin with a little stove in it with a door opening upon the rear of the boat would surely be much more to the point. In two hours, they had reached the first rapids three miles above the refinery. Two of the men threw down their poles and jumped out to push the boat while the other two continued to pole up. Druker held on to his rudder, a long pole flattened at its wide end and turning loose at his will on a stout iron pin. The silence of the weird scenery, the two banks thickly covered with laurels and pines rising hundreds of feet above the river, no sign of habitation anywhere, now and then the shrill cry of a scared whip-poor-will or the mournful hooting of an owl waked up from its slumber by the passage of the boat, all this engrossed the attention of Mark as he thought also, seated upon his trunk, of his loved ones far away, and the strangeness of his new surroundings. Druker stood close to him steering.

"We had a tough time, sir," said he. "Three years ago in winter, when, through ice and snow, on this river and among the hills up yonder, we had to carry day after day those big heavy iron retorts and all the tanks and pipes and machinery for the works. We had however more water than we have now, and the pull was not so heavy. The horses and the oxen had the worst of it."

"You will rest tomorrow, on Sunday?" Mark inquired.

"Rest? No, sir," said he, looking curiously at Mark. "No Sunday for poor folks like us. Mr. Fieldbell don't care much for Sunday. Tomorrow night he will be looking for the load for which we are now going up. He wants it for Monday morning. We will work hard this afternoon to roll down one half of the barrels, and the rest tomorrow till about ten or eleven o'clock."

"Bad enough," said Mark. What miserable management! he reflected sadly. "Surely Col. Albert is not aware of this. How will he feel when he hears of it?" Druker's mind had evidently been stirred up by the question. His frequent looks at Mark while he steered his boat were soon followed by his own inquiry as the rapids were being passed and the boat glided into smooth deeper water.

"What may be your business, Mr. Mark, up this way? I like to meet men who think it right to rest on Sunday. That's the way I was raised in the valley in Virginia. My father would not cut hay on Sunday, even if the weather looked ever so stormy. He had himself been raised on that line by my grandfather who had cleared the big farm on which he died nearly one hundred years old." His voice here began to tremble, and Mark, surprised, could see tears in his eyes.

"How I have come to stray off so low, God knows. But I am tired and ashamed of it. I must break off with all this nastiness and live like a man again, whether Mr. Fieldbell likes it or not."

"This is sound speaking, a good determination, Mr. Druker. I am a friend of the gentlemen in Kentucky who own this boat and these oil works, and I am sure they will approve of your resolve." Thus encouraged, Druker smiled a look of content.

"Let us push hard, boys," said he. "We can, if we will, get to the landing before noon. Then we will have time to roll down our load. We will stay home tomorrow, attend meeting too at the schoolhouse, go to bed early, and at one in the morning, by moonlight, we will start down for the refinery. I am going to have a talk with Mr. Fieldbell about our Sundays."

"Bully for you, Druker," said all the men. "We go in with you."

"Who will preach tomorrow at your schoolhouse?" inquired Mark.

"A Methodist circuit preacher, Mr. Freenberry. He has an appointment there three times a month. A fine man. We all like to hear him; a great Sunday man too. He don't mince his words to us about our Sunday boating, and we don't blame him. I have enough of it for me."

"And so have I," said the four men in unison as they poled and pushed with all their might.

"Well," said Mark. "I am very glad to hear all this. I had been debating with myself whether I could not stay quietly at your house tomorrow, Mr. Druker, instead of tramping a stranger this afternoon to Mill Creek, leaving my trunk with you, and getting to the home of Reynolds at night fall when he don't expect me and know nothing of me. Do the works there run on Sunday too?"

"They charge the retorts every ten hours, and between charges, three firemen keep up the fires. These Black fellows take turns about on Sunday and are pretty sure to have as much of it free as they can. Most of them walk to town on Saturday, spend their extras there on Sunday, that is what extras they have left from pay day on the seventh of the month, and on Monday morning bright early they walk back their ten miles each with his bundle at the end of a stick on his shoulder."

"Will you let me board at your house till Monday morning, Mr. Druker, that is if your wife is also willing?"

"Oh! She will be willing as much as I, if only you can put up with our farm. We are poor folks, but my wife is a neat woman."

"So is mine, Mr. Druker. A good wife is from the Lord."

"A *prudent* wife, sir, for Mr. Freenberry preached to us from that very text a few months ago, and you should have seen mine looking at me all this time. She will be mighty glad when I tell her that I have done now working on Sundays."

It was half past eleven o'clock when they reached the landing, and they could not remember when they had made so short a trip at low water. "Why, hallo!" cried down to them a man from under a shed or the top of the bluff at the foot of which they were pushing the boat to fasten the line to a root. Another man was with him and they were rolling down barrels filled with crude oil from two wagons.

"How early did you start this morning?"

"As usual, about seven o'clock," answered Druker. "But we pulled hard to put out this load and take in another this afternoon and be ready to start back by moonlight on Monday morning. We won't have to pole but moderate to push down, and we will have all Sunday to ourselves. Come down to meeting tomorrow and bring your folks with you, Andrew."

"So I will, I declare, to hear Mr. Freenberry, you mean?"

"Yes, Lewis, you come too," added Druker as he trudged up carrying Mark's trunk.

"Yes, I will," answered Lewis. "We will all come down from our house."

"They could take my trunk now," suggested Mark. "Could they not? And get this note also to Reynolds and tell him that I will be there on Monday in the forenoon?"

"Yes, Andrew," said Druker. "This gentleman is going to stay at our house tomorrow and will go to meeting with us. Tell that to Reynolds and tell him also to come down with his wife and take dinner with us. Now, boys, let us all go and take a bite before we load the boat."

One of the few barrels which were still on wagons were rolled down the skid. Empty ones were rolled up for the return trip to the mine. A small space in front being left vacant to carry the trunk, and the dock men, spurred on by the active, earnest manner and force of Druker, were not slow to start back in time to bring another load of oil to the landing before dark.

"Come along, Mr. Mark," said Druker. "Come to our house and let us have something to eat. My wife has heard our noise down here from our house and must be wondering at the boat being up already. Perhaps she thinks that we want to make up our load before night to be ready for a start tomorrow morning. Won't she smile when she finds that she is mightily mistaken!"

Mark kept up with Druker in his rapid walk along the river road till they reached the bars in his fence, two of which he laid down for them to pass through into a large orchard extending to his log house. It was hidden from the river, behind and below it, by a thick growth of tall bushes, and the over-hanging branches and foliage of majestic old elms. Three large yellow dogs came running and barking from the house to meet their master and licked his hands, while looking somewhat savagely at the stranger. They began at once to smell the scent and lay upon footsteps up to and back from the bars. Two little boys barefooted came also with them as fast as their short legs could carry them, solely wondering who the man with their father could be.

"Your Ma well?" said Druker, taking them up in turn to kiss them.

"Yes, Pa, all right," said the elder. "Molly Pretty had a calf last night, a purt heifer. There they are, see, back of the crib. See how Molly licks it."

"I thought when I left she was not far from it," said Druker. "Well, we must take good care of her. Molly is a good help to us, and so will the heifer be when big enough for that."

"How do you happen to come up so soon, John?" Standing at the open door, his wife inquired, looking surprised also, especially at the companion brought by her husband.

"Oh, well! Tomorrow is Sunday, and we all want to stay home tomorrow, so we pulled up in a hurry, and we will eat something that I may go back as soon as I can and load the boat ready for Monday morning right early. Hereafter I am going to keep on that line, Sally, just as you want me to do, just as my father and grandfather did."

As he said this, Mark observed, as before in the boat, that his voice trembled again with ill-represented emotion while his wife took his hand to press it in hers. Her countenance also beamed with joy. Her heart was full, yet she uttered not a word.

"This gentleman, Sally, is Mr. Mark from Cincinnati. He is going to stay with us till Monday when he wants to go up to Mill Creek. The wagons have just taken his trunk there to Reynolds's house. I see you are frying eggs and bacon for us?"

"Yes, I heard the boat come up and your voices down there at the landing, and somehow I half suspected what you was after, because I know you like to hear Mr. Freenberry."

"Yes, he is a first-rate man—a not afraid preacher. He knows he is right, and he wants to be true to us, mean set that we are. Do you like corn bread, Mr. Mark? Take this chair."

"I do," answered Mark, cutting a piece onto the plate before him, while his quiet thought went to his own circle at home and to the blessing which, at every meal, he prayed for, the daily bread and food prepared by his dear wife. Thus did she and the children think of him also, missing his voice and his presence in the vacant seat at the family board. As he was at his home, so he was everywhere, and so he sat silent for a few moments. Druker and his wife looked at him. She understood more readily perhaps than her husband did, although, as by a flash of memory, the countenance of his father seating down at his meals, seared into his mind with a subdued voice.

"Take some of this butter, Mr. Mark," said Druker. "Sally made it—she knows how. By the by, the children say Molly had a calf last night? It was the first thing the children told me."

"Yes, between nine and ten o'clock I heard her and woke up Abe and Dave to come with me to the stable. I lit the lantern, and we all three went to put some clean hay under her for her calf and help her all we could. She soon had it and began to lick it like a good Molly she is. It's a fine heifer, you saw it, red all over."

"Yes, her first one too, and you will name it for us, Mr. Mark, if you have no objection when you see it, after we are done eating." The meal was soon over and they all went in a body to the crib to take a good look at Molly and Lucky, as Mark named the heifer, a playful little thing, bright and full of life.

"If you come with me now, Mr. Mark," said Druker, "you will see how quick we can unload and load that boat." Entering a footpath through the bushes, Druker led Mark along the high riverbank.

"How many barrels will you put in, Mr. Druker?" he inquired.

"One hundred, all five of us, between now and sundown," answered Druker.

"All this in five hours, down this steep bank?"

"Yes, sir, we can do it." They met at the shed the four men returning from dinner and eating the last of it as they walked in their hurry to be on time. From the boat to the shed above upon the sloped bank, two stout timbers were laid a foot apart, resting upon shortly on this long skid. After the empty barrels had been rolled up and piled in tiers under the shed, the full heavy barrels of oil were slid down lengthwise, the end of each close against the end of another. One man was below, and two turning them off into the boat toward the other men who rolled into three lines, while the other two men in the shed kept the slide always filled with new barrels turned down lengthwise.

"Has the highest water you have known in the Elk," inquired Mark, "ever reached the floor of this shed?"

"No sir," answered Druker. "I have seen it up however within six feet of it after two days of heavy rains."

Mark continued thoughtfully. Two such boats, water-tight and decked over, could be filled here, one after another, through a pipe down this bank, from the contents of an iron tank on wheels. After a few trips, the iron tank itself could be filled from an agitator at the retorts. All this hard labor could thus be dispensed with and saved. The sale of the barrels would cover the cost of the change. One hour before sunset the two wagons, loaded up again, were coming down the hill to Mill Creek. Before they started, however, Druker had inquired from one of his drivers: "Did you give that trunk to Reynolds, Andy?"

"Yes, with the note," he replied addressing Mark. "He read it and read it a long time and did not say much. Only he told one of the Black boys to take the trunk up the hill to his house."

"Will he come tomorrow with his wife to meeting and take dinner with us?"

"Yes, he laughed and said he might, but was not sure. He wondered what took you to that meeting when you knew that Mr. Fieldbell was waiting sharp on Monday morning for your load."

"I'll take care of that," replied Druker. "The boat is most ready now. You see I don't look to him to let me know what I had better do. He needs schooling bad enough himself."

"Well, we will all be down tomorrow to hear Mr. Freenberry. I told the folks all along the creek as I passed by. They all said yes."

"I think we will have a fine day tomorrow," said Druker to Mark. "There is no sign of rain." He was then pushing off the boat and loosening the line to fasten it again after giving it more play over the water. "Now we are all right; let us go to supper and rest. On Monday morning between one and two o'clock, if we live, we will all be here to start down."

"While you were loading," said Mark, "I saw a straggling line of Black people yonder up on that hill, walking down the road toward the Elk, pretty well dressed up. Who were they?"

"This is Saturday, sir," replied Druker. "The Black miners and the Black chargers from Mill Creek wind up their week's work, put on their best clothes, and foot it to town to stay there with their wives and friends, some till tomorrow afternoon, and the rest till Monday morning when they trudge back with new shoes, or new shirts, new drawers, or new pants bundled up on their backs. They get all these from the stores on Reynolds's orders to Mr. Fieldbell who signs them also, or tears them or charges them just as he pleases."

"Col. Albert told me that they pay $150 a year from Christmas to Christmas, for the hire of these men to their master?"

"Yes, sir. The company has also to clothe and feed them in regular rations. The men all sleep in bunks in the two rooms of their log house. Those gentlemen in Cincinnati and Covington have been here but once when the works were being put up, and I take it that they have sent you to investigate a little for them?"

"It is just as you state it, Mr. Druker." Thus conversing together, they had returned to the house and taken seats on the bench fronting the road. Druker went back to take another look at his cow and calf. His children and the dogs all followed him, while his wife was busy cooking supper and setting the table.

"Come, Mr. Mark," said Druker returning. "Let us have supper. I feel mighty hungry, don't you?"

"I have not worked so hard as you have, Mr. Druker, but I feel hungry too, and will do justice to this good corn bread, this sweet butter, and all the good things before us."

The two boys climbed upon the high bench for their seats at the table in the same side with their mother, and when they were all ready, Druker, bowing to Mark, said to him in a reverent tone of voice, "Mr. Mark, will you ask a blessing?"

Mark responded gladly. He almost felt at home with his own wife and little ones. As they were eating, he said to Druker: "Mr. Druker, you have a good crew of men to help you on your boat."

"They are, Mr. Mark, far better ones, than I had last year. Mr. Fieldbell had given me five Black hands whom he had hired also by the year, same as those at the mine and at his refinery, but I could get no work out of them. They were so lazy and full of sleep and told such lies too. Of chickens and turkeys I have plenty now but I could not keep mine as long as these Blacks were about me, dogs or no dogs. The men I have, I picked about here. I pay them one dollar a day. They find themselves, and I am sure the company has not spent since last Christmas for the work of this boat, one half the money they paid last year, while I have boated nearly one half more barrels. So much for us white trash, and so much also for the darkies. Like a good many poor whites, I came over the mountains, from the valley to the licks on Kanawha in order to get away from them and their masters. I don't blame the poor creatures however as much as I do their masters for raising them as they do from the day they are born to be as lazy as they are most of themselves."

His face was flushed with emotion, and he looked intently at Mark whose eyes were also fixed upon him with silent and close attention. He had already thought himself of recommending to Col. Albert to make as early a compromise as could be obtained with the owners, to dismiss as many of the Black hands as had been found unfit for the work in order to fill their places with picked free white labor, as Druker had succeeded in doing. It was therefore with full assent of mind that he replied: "I think on this front just as you do, Mr. Druker."

"I thought you would, and any sensible man will. I don't understand how Mr. Fieldbell can get along with such hands about him, especially for his night work when they are on watch all by themselves and he in his bed far away while they are firing and running these big stills full of boiling oil. What a blaze a fire like that would make of that big refinery! It could be seen all the way from here! A few of them take their nap in daytime and keep awake at night, but mighty few, sir, I can tell you. I know them too well." They had finished supper, and the table had been cleared.

"Who teaches your children in that schoolhouse, Mr. Druker?"

"A young man from Ohio. His six months are just out, and we are trying to raise money enough among us here to keep him all the year, that is ten months, and let him go home for two months of vacation to rest a while. We give him five cents a month for every scholar, and board him free every family one week by turns."

"Is he a good teacher?"

"First rate, sir. Mr. Freenberry thinks that we have had singular good luck in getting him. He is so patient with our young ones, takes such pains with them, and plays with them too. They all like him."

"Here is my first reader," said Abe, the elder child, showing his book to Mark. "Mr. Clark wrote my name in it for fear I lose it."

"And here is my speller," said Dave, holding it up to show it as his brother had done. "You see here where Mr. Clark wrote also my name, David. When he wrote it, he said that David was a King who had been a shepherd when he was a young man, and that, with a sling, he had cast a stone into the forehead of a great big wicked soldier who wanted to kill his brothers, and that stone sank into his forehead so that he fell dead upon his face to the earth. That was in a valley too, between two mountains."

"You remember well all what is told you, my little boy."

"He read it to us too out of the Bible," said Abe, "that same day."

"Yes, he did," added Dave.

"Well, let us read it again," said Mark opening the family Bible which he had seen neatly protected from the dust by a bordered white cloth upon the table against which he was seating. As he read, slowly and feelingly, in the fourteenth chapter of the first book of Samuel, of the intrepidity and heroism of David, all eyes in that room were fastened upon him, and he thought of himself reading to his own loved ones at home, realizing in his mind, as he would have said to them, that David was fearless before man, because he felt and knew that God, the living God, was on his side.

As he closed his reading, Druker's wife said, "Mr. Mark, would you offer prayer for us, with us?" They all knelt, and Mark, with voice at times choked with emotions, asked in prayer that every parent and child in the divine presence should be gracious by given through life the unshaken reliance of David upon the faithfulness and power of a covenant-keeping God to carry they themselves through every city and through every trial of His own appointment.

As they rose, Druker pressed his hand, and said: "Mr. Mark, I am glad you came along with me. Your talk does me good. You are not an old man yet by any means, but you remind me so much of my father and of his ways of thinking and of doing."

"We have only two rooms in our house, Mr. Mark," said Druker's wife. Will you sleep comfortable in this bed, although we cook and wash and do almost everything in this room?"

"It's all right, Mrs. Druker. I will sleep soundly, you may be sure, just as your husband will after so good a day's work."

"Well, good night," said Druker as he went into the other room with his wife and two boys. A quiet, refreshing night's rest was given to everyone, believe me, under that roof. They had prayed for it, as children and a father, and they had asked in vain.

Chapter Four

Day of Rest at the Landing

The next day was a May day, a spring day, bright and cheerful with sunshine, green leaves and blooming flowers. All nature smiled. The breakfast table had been cleared. Druker's wife was dressing her children in their clean Sunday clothes after having given them a good wash. Druker, himself clad in his best, had returned from the shed. He had there looked down upon his boat and found it all right as he had left it the evening before. He was speaking with Mark, who was seated upon the bench outside fronting the orchard, when, looking toward the barn, at the sudden barking of the dogs, they observed an elderly man on horseback raising his hat and saluting them.

"Mr. Freenberry," said Druker. "Come, Mr. Mark, let me make you acquainted with this good man, always on time for his appointments." The dogs seemed to know that the newcomer was no stranger, stopping their bark. They jumped over the fence, and, after thus returning their own friendly salute, moved around the home, wagging their tails, scenting his trail, and looking at the rider.

"Mr. Freenberry, let me introduce to you Mr. Mark from Cincinnati. He stopped here today with us, to attend our meeting, and he believes, as you told us more than once, that Sunday is too good a day to work in and waste. I believe it too now and mean to stick to my belief."

"I am indeed very glad to hear this, Mr. Druker," replied Mr. Freenberry, dismounting. "Glad also to make your acquaintance Mr. Mark," he added, shaking hands with him.

"Let me take your horse, Mr. Freenberry, and you stay with us today won't you?"

"Well, yes, I thank you, although I was on my way to the schoolhouse to see whether Mr. Clark had been kind enough to get the benches and everything else ready for both services, morning and afternoon."

"He has, Mr. Freenberry, for he came yesterday afternoon and told to my wife to tell you so, if you happened to stop here from below." He then took off the saddle pockets and handed them to the preacher who threw them over his shoulder.

"Mr. Clark is always thoughtful. It was very kind of him."

"Come in, Mr. Freenberry," said Druker, lowering the bar. "Have you had breakfast this morning?"

"Yes, I spent the night at Mr. Lewis Young's below. He and his folks will shortly be here." They followed Druker who then replaced the bar and led the horse to the stable to take off the saddle, give the animal some water and make it fast to the rack. Mrs. Druker and her children were at the door to greet Mr. Freenberry as he approached with Mark.

"How is Mrs. Freenberry and the children?" she inquired.

"They were all well, Mrs. Druker," he replied, "when I left them about two weeks ago and I have heard from them since through some of our neighbors whom I have met twice."

"How far do you live, Mr. Freenberry?" inquired Mark.

"About 120 miles from here as the pigeon flies, Mr. Mark, on the headwaters of Campbell Creek." So speaking, he waved his right hand in the direction that a pigeon would take.

"How many appointments have you kept since you left home?"

"Fifteen, I think, weekdays and Sundays counting this," he replied slowly, as if to assure himself that the figure was correct. He entered the house and laid his saddle pockets upon one of the chairs.

"Take this glass of fresh milk, Mr. Freenberry," said Mrs. Druker. "My husband said to me that you just came this morning from Mr. Young's?"

"Yes, I thank you. I have just passed your Molly and her young calf, pretty little thing of life, frisking, bounding, kicking, happy as it could be."

"Yes, sir, she got it night before last. It's a good lift to us. How do you like her milk?"

"Fine, fine," he replied, smacking and wiping his lips.

"See how they are coming from all directions," said Druker, returning from the stable and looking up toward the road and the hills which bordered it.

"Yes," said Mr. Freenberry, consulting his watch. "Half past nine o'clock. It's about time for us also to be going. The bell is tolling."

"Mr. Clark is calling us," he continued. "Who will take the place of this faithful man when he is gone? But I hope he will not go, only for his vacation in the fall."

"Oh!" replied Druker resolutely as he was locking the door of his house. "I will ring that bell while he is gone. I have turned over a new leaf, and feel too well about it, not to go on and turn over many more new leaves." He looked at his wife, while thus delivering himself. She was leading Dave by the hand, as he had hold of Abe's.

"Good words there, Mr. Druker," said Mr. Freenberry. "I am going to preach to you personally this morning, and give you some counsel which, I hope, will help you stand gladly by your honest resolve. I will of course be preaching also to myself and to the rest of us."

"I will be all ears, Mr. Freenberry." Mark promised to himself to be an attentive listener also. The school log house was pleasantly located a short distance off the road under the shade of tall beech trees, and at the mouth of a hollow between two steep hills. The bell hung from a cross piece of split rail fastened with hickory withe to the tops of two long saplings. A large number of men, women, and children, many on horse and the rest afoot, had already gathered. Most of the women with their children were inside, and the men outside securing their horses to the tree or standing in groups talking to each other. More were arriving, among them the four boat men of Druker with their wives and children, and Reynolds also with his wife whom Druker introduced at once to Mark who shook hands with them, while Mrs. Druker and Mrs. Reynolds warmly greeted each other. Clark had just stopped ringing the bell after looking at his watch and ascertaining that it indicated ten o'clock.

Mr. Freenberry introduced him also to Mark, and all went in to take seats in the benches without backs which were soon filled up. The two windows and the door were left open for the congregation outside who could thus hear as well as those inside. After an impressive opening prayer for the divine presence and blessing, which was made still more so by the hearty ejaculation of many an amen from all parts of the room, Mr. Freenberry, holding his pocket Bible in his right hand, invited all to unite with him in singing a well-known hymn, the first two lines of which he gave out, then sang, continuing thus to the end, and, as they had joined him in prayer, so they joined him in song. He evidently felt that he had not come in vain, for, after reading from the book of Acts a portion of the narrative of Paul's shipwreck, he stopped at this declaration of his: "God whose I am and whom I serve" [Acts 27:23]. He emphasized it by yielding to the words of gratitude for his own divine call to the ministry of the Gospel and his love of the work entrusted to him to help his fellow men into and in the way of life out of the slavery, drudgery, and thralldom of sin: " 'Yet not as an eye servant, for eye servants,' said he, 'are but barren drones, men pleasers, ever ready, for sly and treacherous tricks, but as the joyful servant of Christ, doing the will of

God from a ready mind' [Ephesians 6:6]. To such His law is a delight, the Law of His Sabbath for instance, this glorious day of rest, so full of repose, of peace, of joyous anticipations, a day which no man, nor set of men have any right to buy or take from you, no more than you have the right to sell it or give it away. For it is God's own day, and He wants you to consider it and keep it so for his sake and your own sake in this present life which is, after all, not even an hour of time to the life eternal beyond the grave; let me read it to you this law of the Sabbath." And Freenberry read it in the book of Exodus as Moses had received it in the Mount upon the tablet of stone, written by the finger of God: "Such a law written upon your hearts, men, women and children, by the finger of God, and heartily accepted by you, heartily obeyed by you all, will for you indeed to feel and declare like Paul: 'God whose I am and whom I serve.' It will make you free indeed and masters of yourselves, by obeying it to yield your will only to His will, in all your dealings with men."

These earnest declarations from divine authority were addressed to hearers, all of whom labored in the employ of men who, barring few exceptions, professed but little regard for God's law. Mark had, for years, yielded his unreserved assent to it. It was one of the chief rules of his life. Its truth and obligations had always more or less asserted their power over the mind of Druker, and his true wife, before and after their marriage, had never faltered in her endeavors, as a mother also to their children, to assist him in his struggles to live up to the Christian training which he had received in his youth. This influence of this training had inclined him to seek her as a companion for life. Reynolds had also married a good wife, but the love of money and of ease had, long before this, hardened his hearing and his heart against such elevating doctrines and motives of action. Although, as a time server, he deemed it good policy to play the part of respectful listener at church or before his superiors. It was because this divine law was his delight that Clark was the diligent schoolteacher every day of the week, and the efficient assistant of Mr. Freenberry when the days of his appointments came round.

The ancestors of most of the men Mr. Freenberry was addressing had opened small cornfields along the creek and upon the steep sides of the hills, on land which Virginia had, after the war of Independence, legislated to satisfy the claims of her soldiers. Most of their lands had been sold by the sheriff for non-payment of taxes. The present owners, under this dubious title, were but too glad to receive from the squatters their modicums of lease payments, for the specified acres. Besides tending their cornfields, potatoes, beans, and turnips, these men made oak staves in the woods, and hickory poles which they hauled to the salt furnaces on the Kanawha where coopers turned them into salt barrels. Very few of these men and their wives could sign their names, fewer still could read. Lamenting their deficiency, they took pride in their schoolhouse and supported with gladness the manly endeavor of Clark to redeem their children from the bondage of their own ignorance, even if he could not go beyond teaching them spelling, reading, and the first rudiments of Arithmetic, Geography, and History. Ignorant as they were of God, they recognized in Mr. Freenberry a man intent upon doing them good, by teaching to them, with all the plainness he could command, truths which their consciences acknowledged as if divinely heard authority. As the common people heeded gladly the preaching of the Savior, thus did they listen earnest to the preaching of this man, proving their appreciation of his labor for them by the eagerness of their attention, and when the service ended, by the heartiness of their handshakings as he mingled among them.

Some of the more immediate neighbors of Druker would have deemed it a favor to entertain Mr. Freenberry, but they knew where he was staying for the day and felt satisfied that all would be done for his comfort there which they could have devised themselves. As Druker led the way home with Abe at his side, Reynolds and his wife accompanied them in response to his invitation sent the day before. At the table the conversation brought up the subject of the sermon.

"You say, Mr. Freenberry," inquired Reynolds, "that I have no right to sell my Sunday to any man or to any set of men, and that they have no right to buy it from me?"

"Thus says the Lord, Mr. Reynolds."

"Well, I left this morning, as I came away from the works, two Black men to keep up the fire under our one hundred retorts for twelve hours from six to six. These men, each with a shovel and a long iron poker, have to walk all the time of their watch and thus feed and rake fifty separate fires. Their masters are paid for this work by our company, and the men are fed and taken care of when they get sick, also by our company. Were not these fires kept up, the tiles under the retorts and the retorts themselves would cool off, and the twenty-four hours of this long heat would render necessary hard firing during the next twenty-four hours to reestablish the needed temperature for the distillation of our oil. What would you do, were you the owner or manager?"

"If I were the owner or manager, Mr. Reynolds," replied Mr. Freenberry, "I would retain much of the accumulated heat by closing my dampers back of my retorts and would close also all my front openings. I would make less oil. I am aware by lopping off this day's work from your present arrangement may decrease your gain, but might not some improvements be devised in this way by which an increased production of oil could be secured every day of the rest of the week from these very same one hundred retorts? And would not your men, all your men, gladly assist in making such an improvement a success in every respect?"

Mark nodded his assent and looked at Reynolds who was puzzled by this reasoning, and unable to make an answer. His own wife smiled a full assent also to the idea. Mr. Freenberry quietly waited for an answer.

"What improvement do you mean, Mr. Freenberry?"

"Oh, I don't know, Mr. Reynolds, I speak on general principles. I know I would be doing the right thing, and the right thing always pays him who does it."

"I believe that, Mr. Freenberry," said Druker emphatically, "and that's the reason, Reynolds, I determined yesterday morning to reach the landing down here with my boat in time to unload and load it again before sundown, so that I could keep quiet all day today with my men, attend meeting with them and my wife and our children, and go to bed early to make an early start tomorrow morning to be at the refinery unloading my oil even before Mr. Fieldbell comes down from his breakfast."

Reynolds was uneasy. He showed irritation and a dark brow when his wife in approval of Druker's determined manner, said: "Well done, Mr. Druker, do give up all Sunday work. It's mean work."

"When wives stand for the right," said Mark, "the battle is half won. Happy their husbands, happy their children, happy their homes!"

Mr. Freenberry, at the request of Druker, returned thanks for their meal and as he had before prayed for a blessing upon it. They rose from the table, and Reynolds told to his wife: "We had better go home, charging time is not far off." She smiled faintly, but assented, and, getting ready, she kissed Mrs. Druker and the children and shook hands warmly with Mr. Freenberry, Druker, and Mark. To the latter she said, "Mr. Mark, we will be glad to have you with us tomorrow and as long as you stay with us."

"How long will you stay?" inquired Reynolds.

"Indeed, I cannot tell, sir, till I get there. You have read Mr. Fieldbell's note about it?"

"Yes, I have. We will be glad to see you. It is only two miles from here, less than an hour's walk. Why don't you come with us?"

"Mr. Druker and his wife have kindly invited me to spend the Sabbath with them," replied Mark, pleasantly but firmly.

After they had left, Mr. Freenberry looked again at his watch: "It is time also," he said, "that I should be getting ready for my ride on to my next appointment at Fallen Rock. It is nearly three o'clock, and the distance ten miles. I have to be there before seven o'clock, and we have no brother Clark there as we have here, to get the house ready. I have not heard of any sickness among your people here, have you?" he inquired, looking both at Druker and his wife.

"No, sir," answered Druker, "we have not, at least not of any very serious sickness. If there was, we would know from Mr. Clark, for he is not a little of doctor himself. He studies his medicine books at night and is a good nurse to watch after school time when necessary and to prescribe his little sugar pills, homeopathic pills, I think, he calls them. They don't hurt a body, and, indeed, when he can get any sick one to do as he wants him to, he is pretty sure to bring him out all right."

"I know he does," said Mr. Freenberry, "and I can speak from experience. You remember my sore throat two months ago, how long it worried me and how soon I got over it under his advice and treatment?"

"You did so!" answered Druker. "And he was so quiet about it, that only a very few of us heard of it. That's the way he keeps on the lookout for all who get sick around here. He hardly charges them anything and cures them all the same. He seems to love that kind of work."

"So did our Savior. He is indeed a good brother," said Mr. Freenberry, mounting his horse. Bidding them all "God bless you till we meet again," he went up the road on his next errand of mercy.

" 'How beautiful are the feet of them that preach the Gospel of peace' [Romans 10:14–15] and bring glad tidings of good things," thought Mark as he stood a few moments gazing at the good man going away at a quiet pace. He turned to follow Druker and his wife, and they spent the rest of the afternoon with the children singing their familiar hymns which, composed by lovers of God, seal so much of His truth upon the memory and heart of the singer from the cradle to the bed of death, soothing manhood, childhood, and old age by the quiet, joyful contemplation of divine eternal things.

They early retired for another night of rest, and when the clock, in the morning, struck one, Mark awoke and heard Druker dress himself and speak a few words in a low tone of voice to his wife. Then he saw him come into the room where he slept, for the moon shone through the little window.

"Have your men come to the landing, Mr. Druker?" inquired Mark.

"I think they have, Mr. Mark. None of them has sent any word to the contrary. The night is beautiful, we will have no trouble in taking the boat down, and we will be at the refinery between four and five o'clock. When I come back, I may be able to go and see you at the works. Good-bye, Mr. Mark."

"Good-bye, Mr. Druker."

Druker had under his arm a basket well filled by his wife with two days' provisions, for he thought he might have to take a boat load of burning oil from the refinery down the Elk to the river's mouth and up the Kanawha to the steamboat landing which would detain him from home until Tuesday morning. As Druker went out, Mark heard a low bark from the dogs, and soon after some voices down at the landing, then the striking of poles against the boat, and, after a few moments, no sound but the screech now and then of an owl, in one of the tall elms on the river bank, broke the stillness of the night. He was soon asleep again.

Chapter Five

Mismanagement at the Oil Works

As the sun arose Mark awoke again and dressed himself. One hour after, with Mrs. Druker and her children, he had completed his breakfast and was ready to take leave for Mill Creek.

"How much have I to pay you?" said he to Mrs. Druker.

"Oh, nothing, Mr. Mark," she answered with a smile. "Husband and I are but too happy to have had you with us since you came, and the time has passed so swiftly for the pleasure and profit of your visit. Whenever you come this way, be sure to make this house your home. You will always be welcome. I have thought so much of your wife and children. I know she would welcome my husband as I have tried to make you comfortable."

As he pressed feelingly her right hand, he answered: "You judge her rightly, Mrs. Druker. Time will bring you both mothers together to make two fast friends, and our children also as their fathers have so well begun to be." While speaking, he had also taken hold of Abe's and Dave's little hands and was pressing them in his. They stood by their mother, and both little faces were wreathed with smiles as they looked up to him.

A few moments later he was on his way among the hills, following the ruts made along the creek by the wagon wheels from the works to the landing, crossing its repeated windings on heavy logs or planks or boulders where the shallow water ran over a solid bottom of coal slabs extending from bank to bank. Here and there he would observe also long fence rails left high among the first branches by the subsidence of some of the sudden freshets so common in high mountain streams. As he passed along rail fence fronting, from bend to bend, by the side of the creek, some log houses of two or three rooms, his approach would start up the barking of hounds and this warning would bring to the door women and children, several of whom recognized in the stranger a man Druker, the day before, had brought to Mr. Freenberry at the schoolhouse. Some men he would pass by, in the timber, astride three-legged benches, shaving oak staves to be stacked in piles for a first drying, then hauled to the salt licks in the Kanawha, and there sold for making salt barrels. Other men he would see, far up on the steep sides of the hills, at the foot of which stood their log cabins, furrowing their cornfields each with a horse, sometimes a cow pulling a small bull torque plough. The hills rose up abrupt and high on each side of the creek. He met two of the wagons and their teamsters on the way to the landing with their loads of oil.

Soon after they had passed him, he perceived through the trees, in a clearing beyond, a large high frame structure and two tall brick stacks from which issued a thick black smoke. Crossing another bend of the creek, he came in full view of the works, a long row of brick furnaces enclosing the retorts, and, above them, two elevated tramways supported by high timber upon which the coal, in small wagons, was rolled down from two entries in the mine, on the side of the hill in the rear of the works. One end of these wagons was filled by the wheeler at the end of the tramway, and then was thus dumped down upon the floor below opposite the retorts' mouths to be shoveled up into piles for charging these retorts when the regular time arrived for opening, raking, and charging them again.

He stood a while under a tree surveying the scene, unperceived by anyone. On his left, midway up the hull side, he saw the double log cabin which, he had been told, was occupied by Reynolds. A small vegetable garden around it was fenced in with it, by long rails laid one on another and held tight by riders, Virginia fashion in the same hill. Below this cabin stood two other ones, also double, which he recognized as the quarters in which the Black hands cooked, ate, and slept, their bunks framed one above another in two tiers against the sides of the cabins. To these bunks, in summer, they would prefer to sleep upon the floor, wrapt up in their blankets before the large grate in which they kept a cannel coal fire burning day and night almost all the year round, except during the hottest days of summer. Beyond the works, near one of the entries, he could also see the double cabin occupied by the family of the head miner, a white man.

It was nearly ten o'clock when the Black chargers who had shoveled up their piles of coal ready for the evening charge were sitting down near them upon their heels, two or three together talking, while the firemen were slowly going from one fire door to another, raking down ash from the grate bars and shoveling in coke left piled against the brickwork of the furnace from the cart raking of the retorts. Along the creek huge heaps of the same coke were smoking and burning to ashes, having been wheeled off by the chargers and of their way before opening their retorts in the morning at six o'clock, a performance which they would have to go through again before charging at it in the evening.

"Mr. Mark, good morning. I was looking for you." Mark turned himself in the direction where the voice addressed him in loud tones, and saw Reynolds standing upon the narrow platform in front of the cabin looking down toward him.

"Good morning," Mark answered, stepping forward.

"Come up around these two cabins, Mr. Mark," continued Reynolds, going down the few steps of his platform porch, and below them the path down the hillside to the works. His wife soon also appeared at the same door and greeted the eye of Mark with a smile and a bow. Some of the Blacks, hearing the voices, had, like Mrs. Reynolds, come to the doors of both cabins to see who the speakers were, and soon all eyes in the works and above the works on the hillsides, were turned to the spot where Mark and Reynolds were shaking hands together.

"I thought you would be here about this time," said Reynolds. "Come up to our cabin. My wife had fixed up a bed for you as well as she could, and your trunk is there all safe and sound. You had no trouble in finding the way?"

"No. It was plain enough along the creek and across it. I followed the ruts of the wheels." Mrs. Reynolds waited till they came up and gladly shook hands with Mark.

"Take a seat," she said. "You must feel a little tired after two miles of such a rough road and too many crossings on those logs. I had another taste of them yesterday both ways."

"Oh, I am all right," answered Mark good humoredly. "I had a pleasant walk as I had a pleasant boat ride on Saturday up the Elk with friend Druker. Did not all this wild scenery look odd to you when you came up the first time from Cincinnati?"

"It did, sir, but, mind, it was in winter, and I had to bring all our furniture along, for my husband was already here and could not possibly that day leave the works to come down for me. I had a time of it through the ice and the snow in the river, in the road, and through the creek."

"You are brave, Mrs. Reynolds, but you knew that your husband needed your help, and you felt the stronger from the hardships of the way."

"That's so, sir," she answered, looking at Reynolds.

"Yes," added Reynolds, "she has always been a great help to me ever since we were married."

Mark took a seat on the platform with Reynolds. His wife went out of the back door to a modest frame kitchen, which, with two little low sheds, put up on the same line for sheltering a few pigs and chickens, made up, in the rear of the cabin, one side of a small potato patch on the rise of the hill. The cabin itself and a short pale fence closed the patch on the lower side, while the remaining two sides were also fenced in by flat pointed stakes nailed on two cross pieces of narrow boards which were themselves fastened upon pieces of scantlings buried in the ground.

Looking down upon the smoking heaps of coke which had been wheeled off along the creek, Mark made this remark to Reynolds: "Much of this coal is hardly coked, Mr. Reynolds. See how it burns and smokes. There is a great deal of oil in it yet."

"Yes, 'tis hard to make them firemen do their work as they ought to."

"Twelve hours coking?"

"Yes," replied Reynolds uneasily.

"How many bushels of coal to the retort?"

"Four bushels."

"Do these fill the retort completely from end to end?"

"They do. That's the standing order from Mr. Fieldbell, with twelve hours' coking."

"When this coking is well done, is there any oil left in the coal?"

"There ought not to be any, he says, but we cannot get it all out so dry as that. There will always be some left uncoked, fire as we may."

"Could not this loss be remedied by charging less coal?"

"It might be."

"Two or three hundred bushels a day saved would make up quite an item of gain in six or twelve months," said Mark.

"It would indeed; sir, but I have no order for that. Mr. Fieldbell would get scared at the idea of getting less oil than he does now."

"It might be tried on ten of the retorts and their condenser."

"Yes," replied Reynolds. "There could be no objection to that."

"Would you let me have a quart of your crude oil, Mr. Reynolds, for a test? I have some instruments and chemicals here in my trunk for this test."

"Yes, sir, we will take this pitcher along, if it is large enough?"

"Yes, it is. Half full will do." It was past eleven o'clock. The charges had been coking over five hours, and the lids of the retorts were all smoking badly, showing indifferent luting or careless scraving up as they passed along the front of the retorts. Mark took out here and there an iron plug off the fire front, and looking in through the brick wall, could see through the bricks by the color of the sides of the retorts, and by the accumulation of ashes at their rear in the line of the draught that the firing needed as much improvement as the lids. He felt pained at these sad evidences of mismanagement, but said not a word, although he could observe confusion on the countenance of Reynolds who was watching all his movements.

"Have you any way to measure these four-bushels piles for the retorts, Mr. Reynolds?" inquired Mark.

"Well, they guess at it pretty high. They are so used to it," replied Reynolds.

"A square light frame without bottom to each man to measure his four bushels for each charge would tell him the rate without mistake."

"That's true, sir. It would be a good trick."

"You would thus know that the four bushels have actually been charged into each retort and this would assist you also in figuring up your receipts from the mine."

"It would indeed, but Mr. Fieldbell counts that himself, I at least think he does."

What looseness in all this! thought Mark. He could also see, from the light pressure upon the gas lights over the retorts, from the condenser, how little of the oil was being distilled. Some of his remarks to Reynolds had been overheard by one of the firemen. He had passed the word along, and their sudden activity at the fire fronts was a painful evidence to the observing eye of Mark and of Reynolds as well, that they knew well wherein they lacked. But this momentary stir could not hide the natural indolence of their movements, the careless handling of their tools, nor their superficial cleaning and feeding of the fires. Their torn shoes also and the tied rags which covered their bodies and their legs told, in part, another sad tale of the utter neglect in which they were left to shift for themselves, of the utter disregard also of the express terms of their hire which called for decent clothing as well as wholesome food and care in sickness. These men work as they are clad, thought again Mark, and I fear, as they are fed, and perhaps as they are cared for when sick. I wonder how often, from Christmas to Christmas, their masters come up these ten miles to investigate these things and satisfy themselves as they ought?

Reynolds led Mark to the large sunken tanks into which the crude oil was constantly flowing down from the condensers, and from which it was pumped into similar elevated tanks at the bottom of which the water and the tar cleared settled down to be run out, before the oil was let down into the barrels. He dipped the small pitcher which he had brought down into one of these low tanks, and, at the request of Mark, poured its contents into a half gallon glass bottle which was at hand upon a shelf. As he handed it to Mark, the whistle blew for twelve o'clock.

"Dinner time, Mr. Mark. Let us go up. My wife is waiting."

The Black hands were also on their way to their meals in cabins from all parts of the works and of the mine.

"How soon do you expect the wagons back from the landing, Mr. Reynolds?" asked Mark.

"In about an hour. I have one load ready. As soon as we have had dinner, I must come down again and pull more barrels to make up the other load."

Mark thought again of his iron tank on wheels into which the oil could so easily be run down from the elevated one, were this one raised a few feet more. So much time and labor could again be saved here; and, best of all, the refinery should have been, from the beginning, built on this spot. Till it could be moved, however, the cost of this present transportation could be reduced very materially to the advantage of the company, and the well-being of the laborers especially, by adopting the process which he had introduced at Newark.

As they reached the house, he laid his sample of crude oil upon a shelf and looked at it carefully for any water that might yet settle down from it to the bottom.

"What do you think of the works, Mr. Mark?" inquired Reynolds's wife, looking steadily at him.

"Well, Mrs. Reynolds, we are all of us so imperfect that the most willing to do right find it hard work to do it well."

"That's so, sir," she answered glaring at her husband, who silently and sullenly took his seat at the table while showing his seat also to Mark. He had already realized the superiority of his guest's moral character and practical attainments not only over his own, but over Fieldbell's also. He clung to his nest as Fieldbell had called it: how should he retain his hold upon it? was the crafty question now running through his mind. He cared but little whether his methods were advantageous to his employer or well-liked by the men under him. The love of self and of unlawful gain must be gratified and sustained, cost what it may. How could it?

"Mr. Mark," he inquired, as they were eating. "What do you think of Mr. Fieldbell's arrangements on the whole down at the refinery?"

"Well, Mr. Reynolds, my answer just now to your wife might apply to your question also, and Mr. Fieldbell would doubtless say yea to it, as I would if I were in his place or yours. We are bound to learn something new every day to improve our ways. This world moves, so must we—if we don't we get left, as the saying is."

"That's so, sir, and I must move down now in a hurry to fill my barrels. You are going to try this oil, are you not?"

"Yes," replied Mark, turning to his trunk to open it.

"Are your chemicals the same that Mr. Fieldbell does use for his refining?"

"I think not, although I presume he knows of mine, if he has not the facilities to prepare one of them."

"Which one?"

"The alkali."

"The what?"

"Alkali."

"What is that? I know he uses vitriol acid then something else. Soda, I think."

"Well, soda is an alkali, but there are several compounds of which it is the basis, all differing from one another in their action, to make soap for instance, or to refine coal oil."

"Did you ask him what kind he does use?"

"No, I did not. The French say '*chacun a son gout*.' We have all our preferences."

"Don't you attach much importance to your own alkali, Mr. Mark?"

"I do, and I will show you in this sample, when you come back, that oil can be purged of its tar without passing through the still."

A look of astonishment spread over the features of Reynolds as he said: "But he distilled it first to get rid of the tar before he uses any chemicals. Don't you do the same?"

"No, I don't. I take a shorter cut, save time and money, and get a finer oil, using but one still, instead of two and three."

Reynolds still wondering, while his wife was listening, made no answer, but turned and went down his path. I must see Fieldbell, said he to himself. He weighs but a feather against this man, and he don't know it. But they will soon find it out down the river at Cincinnati, and if he don't stir the mud to hide, he is gone sure, as well as myself. I must go down tomorrow and see him. The two wagons, for whose return from the landing he was looking, were in, right below the first crossing. He saw them and he hastened to the tank to make up the load.

"Is that man with whom I saw you talking from the tank the one you were looking for, Reynolds?" inquired the head miner.

"Yes, the very one, and a smart fellow he is too, I can tell you. You had better look out, Johnson."

"Look out, for what?"

"Well your coal digging."

"What about my coal digging?"

"Well your deliveries. He says they ought to be measured every one of them."

"He is right. I know what my diggers send out every day eighty bushels each, and when there are extra bushels, they send out more. They hand you tickets for their extras every Saturday morning, and you give them orders on Mr. Fieldbell for them."

"That's all true." They looked at each other.

"How does he say that you ought to measure them?"

"Why, by measuring every charge that goes into the retort."

"How would he have them measured?"

"With a square frame, without bottom, for every charger."

"That's first rate. He is right, and if it was so, you could tell me which wheelers cheat, for I cannot look at every load that comes out. The entries are so deep, there are so many rooms, and I have so many pillars to put in every day."

During this conversation, Reynolds was filling and burying his barrels. His manner was distracted. A Black man, under his directions, was firing the boiler and running the engine to pump the crude oil from the lower tanks into the elevated ones. The second wagon was also being loaded.

"That's a fine man who came this morning," said one of the teamsters. "We met him on the road, and Mr. Druker speaks mighty well of him."

"Yes, he knows a good deal," answered Reynolds. "He says he can teach a great many things to Mr. Fieldbell about refining coal oil, beat him all hollow."

"Can he indeed?" inquired Johnson.

"That's what he says," replied Johnson. "He can make better oil, and cheaper too."

"Did he tell you how?"

"Well, some. I don't understand it. But, to begin with, he don't clap this black oil into the stills to throw down the tar as Mr. Fieldbell does. He has a shorter way to do that with some kind of soda. He is working at it now at our house upon a quart of oil which he took up in a bottle before dinner. Do you see him at the door, shaking that bottle?"

"Yes, he does," replied Johnson. They all looked up from behind the large engine, and could observe his motions, unseen by him. Having vigorously shaken the bottle with both his hands, he was now examining through the white glass the settling of the contents in the bottom of it. "I'll go on with this work and fill up the barrels for the wagons tomorrow," said Reynolds. "I want to go down to the refinery in the morning, Johnson, and you will look after the teamsters, that they get their loads morning and afternoon."

"What takes you to the refinery so sudden?"

"Well, I want to have a talk with Mr. Fieldbell about this stranger."

"Don't you like him?"

"I can't say I do. Mr. Fieldbell writes to me that he has been sent from Cincinnati to look us over, and he is certainly sharp and smooth as a razor. He cuts to the quick and the bone. I must have an understanding about him with Mr. Fieldbell. He has just as much at stake as I have."

Johnson was an honest miner and did not entertain for Reynolds any more regard than the Black hands. Some of these in the works who had overheard the conversation before the two men were repeating it along, and soon the diggers in the bank knew all about it, as a result already the fires were being far better fed and cleaned than they had been in the forenoon. The coal wagons, when wheeled down from the mine, but before they were turned at one end for the dump down, were stopped a moment by the wheelers who would then look up toward Reynolds's house to catch a glimpse of the Cincinnati man, who had said that all the coal should be measured before being charged. Their motions seemed to tell him: "Here is my load full to the top, see." But they knew that he was right: Mark's ways were not overbearing, pleasant rather, and sensible, not rude and coarse like Reynolds's!

All these movements did not pass unobserved by him, nor by Johnson. The miner shouldered a new post and climbed up one of the wheelways to lay it on one of the empty wagons which he followed into the bank, determined in his mind to keep up his wheelers to the mark, and Reynolds lost no time in preparing the four loads of full barrels for the next day. When the clock which hung on one of the tall poplar posts that supported the roof struck four, he had nearly completed the work, and the chargers had commenced wheeling off the greater part of the coke which they had left the previous morning to fire the retorts. He saw Mark making his way down from the house to the works. A rapid glance over the piles of coal shoveled in the morning opposite each retort satisfied him apparently that but few of them would fill the retorts as four bushels should, and he sent word by the Black hand who assisted him on the engine, to all the chargers in fact, that they must shovel up their piles. He was sure to make full charges. They all grinned more or less when they received the word, and added as much as they thought would do, then went on wheeling off their coke, each man having ten retorts to clear and to charge.

Mark noticed all, while walking quietly through the works. He approached Reynolds as he was filling the last barrel: "Why, Mr. Reynolds, do you this every day to prepare your load for the next?"

"No, sir, but I am going down to the refinery tomorrow morning, and these barrels will be ready for the teams so that they can go on making their regular trips."

Mark suspected the object of this sudden move, but, without seeming to attach any weight to it, he added: "I have just requested your wife to be kind enough to keep some supper for me. I told her that I wanted to stay down here till this charge is well over. I want to see it all."

"That's all right, Mr. Mark. We take supper as is, but I will go up only for a few moments to eat some and will come down to be with you again till they have all done."

Strange, thought Mark, that his chargers eat their meal before they come down to prepare for work, and that he leaves them to go up to his when they should know that he is down here watching them closely. But Reynolds did not seem to realize the incongruity of the thing, not to stigmatize it more harshly. Although he could probably call his house on the hill his eyrie, even if, from it, only one half of the retorts could be seen and observed, as the sides of the works were not boarded up.

"Have you tested that oil, Mr. Mark?" he inquired.

"Yes."

"What do you think of it?"

"It is heavier than the oil at Newark. It has more tar in it, but it can all be thrown down. In both yields alike you can see it in the bottle on the shelf when you go up."

"All the odor gone from it too?"

"Yes, it is completely deodorized."

"How much tar have you thrown off?"

"About one third."

"One third! Why, that much then could be thrown off here before the oil is carried to the landing?"

"Yes, one third less to carry, and that much less to worry the refining."

"What would you do with all this tar here?"

"Burn it every time under the retorts instead of coke."

"How?"

"In a stout trough made of stone fired clay here and laid upon your grate bars, it could be fed by an iron syphon provided with an iron funnel outside against the front wall over the fire door. There would be no tar then to run into that deep dry well which has been dug in the clover field close to the refinery, from where it must taint the water in all the wells around."

"It does. Everybody complains of it down in the Elk valley."

"But, with the tar off here, what of the refuse and waste oil from the agitator at the refinery?"

"It should be burnt also in a clay trough under one of the stills, and kept from running out into that well."

"An agitator would be needed here then to throw down this tar?"

"Yes, but I want to go and see this charging." Mark thought he had said enough. He knew it would be repeated the next day to Fieldbell who would try to make the most of it to his own credit in some way by letter to Cincinnati. But his own letter would reach Col. Albert and Col. Truesdell before Fieldbell's would, and in it he would report the result in part of his investigations at Mill Creek, as he had reported in his first letter his impressions of Fieldbell and his work. He had also come to the conclusion that, in order to assure the success of the improved process which he would recommend to them, his own post of duty should be at Mill Creek. Fieldbell should remain at the refinery, but Col. Albert should arrange to come to the Kanawha himself and take there, as treasurer, personal directions of the affairs and interests of the company.

The chargers had now cleared off the front of the retorts all the remaining coke, and were busy luting with mortar the edges of the fresh lids intended to close the mouths of the retorts after they filled them with fresh charges. Seeing the stranger walking around, and looking at them while thus engaged, after Reynolds had left him to go up to his house, they seemed to pay particular attention to the tempering of the luting, and its being free from any pieces of gravel through which the very hot vapors would find and force a passage out of the retorts and be lost. The clay given to them for this mortar had not been carefully ground and their loud and expressive opinions of the man in charge of this particular work were far from being complimentary, for they had, with their fingers' ends, to search leakage and pick out all such causes of leakage, and they felt pretty certain that they could not find them all.

As soon as the clock struck six, the ten chargers unscrewed the hot lid of the first retort on each bench and threw it back out of their way, their hands being protected by wet bagging. Ten high blazes rushed up out of the ten open mouths. With long iron prongs and rakes, they pulled out the red-hot coke. As it fell upon the brick pavement, they threw buckets of water from full barrels at hand which were fed by pipes from large cisterns extending under the roof. The hot steam ascending from the red coke on the floor would put out now and finally the oily frames above the mouths of the retorts would be doused. In a few minutes the coke would all be pulled out, and the interior of the retorts, eight feet in length, one foot in height and shaped like inverted U's would present a red-hot cavity into which the black lumps of coal were rapidly shoveled up. They were first pushed up with the rakes until each pile was thus emptied off. When the baked old mortar still adhered to the mouth, it was hurriedly scraped off with the cutting edge of the shovel, and the fresh lid was clapped on, then screwed with due consideration to the observing eye of the stranger, who, they all knew well, was not to be driven off by the smoke, the steam, the black dust, and the roaring bustle around him.

They did not care for Reynolds up at his supper, but they did care for this man, even if he was but a stranger to them because he was there in sympathy with them, helping them by his presence. Had the luting clay, they felt, been calcined and ground under his direction, it would have been fine and smooth like putty, pleasant to handle, free from all gravel. A careful inspection of the charges in the retorts satisfied Mark that wherever the first shovelfuls of coal had been pushed back with the rake, a sufficient empty space of one foot or more remained in front to receive more coal had the pile been larger. Where the rake had not been used, the front of the retort was crowded up almost to the mouth. He was satisfied also that his remarks to Reynolds about not charging these retorts to their full capacity was correct, unless sufficient heat for a perfect coking and distillation of the whole charge were maintained during the twelve hours.

Such were his thoughts, when, in the midst of all the noise, as he stood back observing the rapid movements of the chargers, he heard the voice of Reynolds addressing him: "That's a very fine test you have made up there, Mr. Mark, an easy oil now to finish with perhaps only one distillation instead of two or three."

"Have you smelled it?"

"Yes, there is no more any odor about it—it's a fine process. What do you think of this charging?"

"Pretty well done, but your men grumble a great deal about the gravel in the lute."

"Yes, that's Wag's fault, and the fellow has nothing else to do under that shaft but to keep his baud well laced up and to shovel back his calcine into the mill as long as it is not fine as flour, but he is lazy like a louse."

"Perhaps the clay was not sufficiently baked?"

"Well, that's his fault too. He has nothing else to do."

"Has he to wheel it from the bank?"

"Yes, from that bank back of our house. He sat on his wheelbarrow looking at you ever so long this afternoon, while you was making your test. You did not see him?"

"Yes, I saw him. At what time will you go tomorrow morning?"

"Right after breakfast, about seven o'clock."

"When they will be charging here again?"

"Yes, sir. It's a three-hour ride and three hours to come back. Three hours more to do some shopping and see Mr. Fieldbell, will make up nine, so that it will be about four or five o'clock when I will get back."

"Who will be in charge during your absence?"

"Johnson, the boss of the mine. He knows all about the work."

"The engine also to pump the water and the oil?"

"Yes, we have been here together these two years."

"You need steady firing, Mr. Reynolds, to full coke these charges as they should be."

"Yes, I know it. It's an important point."

"Have you tried all your hands at it to pick out the most reliable for this work?"

"Well, no, not all. Our best men are diggers and have been in the bank since they were first hired two years ago. Johnson picked them out for himself."

"If your firemen don't do, could you not get some white hands about here in their place?"

"Yes, perhaps so, just as Druker did last Christmas. But white men require watching just as much as the Blacks." And a poor watcher you are, poor fellow, thought Mark, as he went on to see the charging of the last retorts, and Reynolds to his engine, then up to the water tanks to see how much more water was needed in them.

Soon the chargers began to wheel away toward the creek the coke not wanted for firing. The night had come, and, in the dark, along a narrow path they had to trudge their way among the burning heaps. The wheelbarrows were worn and burnt, in a sad condition for this work, as poorly filled as the wheelers in their shoes also torn and burnt. Reynolds's poor management was written everywhere. He saw from where he was that Mark was turning his steps to go to the house and he came at once to join him.

"You must feel hungry by this time, Mr. Mark. My wife has kept some supper for you."

"I thank you. I will do justice to it."

"How long do you think will you stay with us?"

"A week or two, that is if you can accommodate me?"

"Oh, yes, if you can get along upon our fare."

Reynolds's wife was at the door looking for them.

"All ready for you, Mr. Mark," she said, showing to him the little table set with his plate and a few dishes. An oil lamp threw a pretty light upon all things around in the room.

"I thank you, Mrs. Reynolds, for all this trouble, but I felt very desirous to see the whole of this retort charging. Indeed, I do not want to miss any. How will that do?" he inquired, looking at Reynolds.

"Mr. Mark is to stay with us a week or two," said Reynolds to his wife.

"Oh! Well, we can have breakfast after charging time in the morning, and supper before it in the evening."

"Yes, yes," said Reynolds, "as well as not."

"I thank you again, Mrs. Reynolds," said Mark. He turned to his sample of oil on the shelf, held it in his hand, between his eyes and the light of the lamp.

"What a clear green color that is!" exclaimed Reynolds, "and there is not particle of tar in the bottom of the bottle, just like the paraffin lubricating oil down at the refinery after three or four distillations and as many treatments. Could it burn in this lamp as it is?"

"Yes," said Mark, "but it would clog the wick in two or three hours, although it would last longer than the lighted oil in this lamp which has passed through the still. The light oil in this sample has also to be distilled to separate it from the paraffin oil." He sat down to his supper and paused a few moments reverently in prayer. Reynolds looked at his wife upon whose countenance appeared almost simultaneously the same devout feeling of thanksgiving.

"Mr. Mark," said Reynolds, after a few moments, "can I take this bottle with me tomorrow morning to show it to Mr. Fieldbell?"

"Oh, yes," replied Mark. "And you repeat to him all I have told you about it. Bring it back with you however."

"I will," said Reynolds. Such open dealing dazed him. In those days of high-priced processes for this new industry in which so much capital was being readily invested, especially at the West, while the Pennsylvania petroleum was inviting and attracting it in the East. Mark had it in his mind however not only to win the confidence and respect of those who had sent him, but to command also the same respect in those who unconsciously felt that they had failed to live up to the responsibilities of their trust, and that they had even willfully deceived their employer. This he knew that Fieldbell and Reynolds had actually done, for the evidences were before him.

The presence of Col. Albert at the refinery would naturally cause the prompt return of Reynolds to Cincinnati and compel Fieldbell to recognize the vigilance of an authority which could not brook imposition but would in the spot require full and good work for full and good pay. Mark knew that Druker would leave the landing again on Wednesday morning at noon. On the next day, Tuesday, he would hand to him a letter to be mailed by himself at the Court House to Col.

Albert, and, with it, another sample of refined crude oil which he would request him to deliver personally also to the clerk of the *Ellen Fray* or to the clerk of the *Annie Laurie*, either of whom, on the return of the boat, would bring the receipt for it signed by the Colonel. By Druker also he would write a second letter to his wife, to be mailed at the Court House.

The evening air was pleasant, and, through the open door in front, the sound of a banjo could be heard plainly coming up from the Black hands' cabins below, accompanied by the singing, the laughter, and the quick steps of a Negro ballad.

"This is Pole," said Reynolds. "One of the wheelers, the musician and clown of the troop. We call him Pole for short, but his name is Napoleon. He is our fastest wheeler, and a good liver. He eats up all his extras."

"You give them rations?"

"Yes, sir, seven pounds of pickled pork or corn beef per week, half a bushel of cornmeal, one quart of molasses, and one pound of coffee, all costing about one dollar and half."

"And his extras consist of what?"

"He is paid for all the coal which he wheels beyond his task per day, and buys with it every week when he goes to town, some sugar, fish, flour, beans, potatoes, and the like which he brings here to cook with his rations."

"Do they all do the like with their extras?"

"Yes, more or less."

"You pay them these extras?"

"Yes, Mr. Fieldbell sends me the money, or else I give them an order on him for such things as they want from the stores we deal with."

"You keep an account for each man?"

"Oh, yes, we settle every week. Each man has his tickets for digging or wheeling, or fixing."

"Tickets, you say?"

"Yes, these tin tickets," and he pointed at several piles of thin sheet iron labels laying on the table, roughly cut with a cold chisel, each pile of the same shape and wired together by itself.

"You do not keep a record then for each man?"

"Oh no, I settle every week with them."

I wonder how Fieldbell keeps his accounts at this rate, thought Mark. Reynolds's wife was clearing the supper table.

"Mr. Mark," she said. "As my husband goes to town early tomorrow morning, we will have breakfast at five o'clock before charging time, same as Mrs. Johnson does." And with these words quickly spoken she glanced at her husband.

"Yes, Mrs. Reynolds," answered Mark, "that will suit me very well." Looking down towards the works, he added: "I will go and take a look at these fires, Mr. Reynolds."

"Very well, Mr. Mark," said Reynolds. "When you come back, here is your bed all ready for you."

Nodding his thanks to Mrs. Reynolds, Mark went out. As he passed the two cabins occupied by the Black hands, and brightly lighted by the large lumps of cannel coal burning in the wide grates, he could see from the good humor which prevailed among them that, taken as a whole they seemed to be an intelligent-looking set of men, and, fairly treated and managed, they could be made to yield satisfactory labor as miners and wheelers, for these had all be chosen by Johnson, and even as chargers. He wanted now to satisfy himself about the firemen, those at least who were on watch at night, and, in order to avoid being discovered by them when the moon would rise, he rapidly walked along the creek until he reached a dark spot under one of the wheelways, where he sat unseen over an hour. From there he could obtain a good view of all their movements, and he smiled repeatedly as he saw them looking up to Reynolds's cabins, while actively engaged each on his round of duties. It reminded him of the looks of the wheelers in the afternoon, which he had also observed while he was making his trial of the crude oil. He felt rewarded for what passing hints he had given them in order to obtain a fair coking of their charges, and came to the conclusion that if the Black hands of Druker had not done well with him, these, under good and steady control, at a work which was not perhaps as exacting as boating oil on the Elk River, could be done most probably far better than under the clumsy hand of Reynolds.

I must now go up and write my two letters, he thought. The moon was rising, and, taking the same path along the creek, he made his way back toward the Blacks' cabins. They were silent now though still bright with the light from the grates, and he could hear plainly the many snores of the sleepers as they had richly earned the sound sleep which had evidently fallen upon every one of them. The doors were wide open and he could see the floor well covered with their black forms wrapt in gray blankets. Walking up the path, he soon found himself at Reynolds's door, and saw the latch string hanging outside. He pulled it to enter. Reynolds and his wife had both retired to their room, and, with the aid of the lamp which they had left burning, he soon commenced with his wife, and continued with Col. Albert, a converse which kept him deeply interested until about midnight. He opened his mind to both, as if both had been there with him, and he felt so confident of the soundness of his advice to Col. Albert that he could not but believe he would, in accordance with it, come up to the Kanawha and assist him in rescuing a fine property from the danger which threatened it, even if he and his little family would have, on this isolated spot, to bear the brunt of this rescue. Fieldbell would have for his share the comforts of civilized life in the Kanawha and the pleasurable company of the worthy Colonel and perhaps of his family.

He was sealing his last letter when he heard a soft stepping outside, as if of bare feet on the ground below the door. Opening it quietly, he looked out and saw by the clear light of the moon, which illuminated every object below and facing the cabin, one of the ragged firemen carrying an empty pail in his hand. His hair was thickly streaked with gray.

"What is it you want, old man?" he inquired.

"I come up to fill our pail at Mr. Reynolds's spring, sir, back on the hill," he answered, somewhat staggered at the sudden appearance and calm manner of Mark, "and I did not know but some of you was sick by seeing this light of yours burning so late. Excuse me, sir, if I have disturbed you."

"Oh no, you have not. I have been writing all the evening." And continuing with a low voice: "How does your coking get on?"

"First rate, sir. We will have a good batch of oil tomorrow morning. You will see for yourself."

"Well, well, that's all right. Good life, good conscience. What's your name?"

"Old Tom, sir. Can I go to the spring?"

"Yes, yes, Mr. Reynolds is in bed, but you have hot work down there, and I suppose he don't object to your wanting a cool drink, every one of you."

"There is another spring down that bank over the creek, but not so cool as this, and don't taste near so good. Coppers are in it, or alum, something of that sort."

"Well, Mr. Reynolds knows all that. You go on and fill your bucket."

"Thank you, sir, please give us your name."

Give *us*, thought Mark, the poor fellows. "My name is Mark, Tom."

"Well, good night, Mr. Mark."

"Good night, Tom." And Tom went on while Mark stood at the door admiring the beauty of the scene, the tall tree, which covered to its top the rugged sides of the hill on the other side of the creek, whose bend at its foot formed the extended flat upon which the works had been erected below the coal bank. The fall of one of those old black oaks had, three years before, revealed among its roots the presence of a vein of cannel coal which had remained so long hidden in the bowels of that ridge; and this discovery had led to the purchase of the two thousand acres by the Kentucky company from an old Virginia family who had bought the land at a tax sale and felt more interested in enlarging with this money their salt works on the Kanawha than in venturing for extracting oil out of their coal. They had been anxious that their Kentucky friends should succeed, for they had still more of the same land on that creek to sell for which they hoped to obtain a higher price, the bed of the bank of the creek itself being a bed of peacock bituminous coal.

How will Sarah like this life? thought Mark. And our children, how can we bring them up here successfully through their tender years without any help from school, without the greater help from the ordinances of God's own house? Will not their young minds remain stunted from such close contact now and for several years with the ignorant and wild youths reared upon these hills? But then do they not fare still worse who cheerfully leave loved and refined homes to go to lands beyond the seas and share their knowledge and joys of the Gospel with millions whose language they first have to learn and whose modes of living must for a long time prove so great a trial to their health and the habits of their own early life? Why should we not ourselves share with these poor ignorant people all the better knowledge which we have received as a trust? And have I not been providentially brought to this very spot for a purpose?

Such were the thoughts which crowded upon his mind, as he stood yet at the door when old Tom came by again carrying his pail filled this time with cool water.

"Good night again to you, Mr. Mark," he said.

"Good night, Tom." He turned back, shut the door, knelt in prayer for continued protection and guidance as he knew that his dear wife and their children with her had done a few hours before, and then laid himself down to rest. The men at work below, for the rest of the night, saw that a dim light was now burning where it had shone so bright until then, but they continued active as they had begun. Tom had told them his tale of the stranger, and one of them had spied his form when he had risen from his seat on the stump under the wheelway to return to the house. They felt that his eye was upon them and they felt rightly, for, at four o'clock, when Mrs. Reynolds awoke, Mark was up already well refreshed by his sleep and going down the path to the works.

"Good morning, sir," said to him some of the hands in the two cabins below who were cooking their breakfast before their large grate.

"Getting ready for work?" he answered pleasantly as he passed along, the moon still shining bright above his head.

"Yes, sir, will soon be there."

The gas lights blazing high over the retorts indicated a strong pressure in the condensers and an abundant yield of oil from the coking of the last charge. The small heaps of coke remaining from the firing of the night showed also that the men had drawn heavily upon them to help their work. The absence of gas jets around the retorts' lids spoke too already a notable improvement in the baking and grinding of the lute clay. Mark was highly pleased with the look of things, and commended, in his own mind, the good judgment of Johnson who had had the most to do with the original pick of these hands, being a Virginian miner himself.

He looked up toward his house and saw him standing at his door. Their eyes met and they greeted each other. The smoke going up out of his low chimney and the busy movement of his wife showed that she was preparing his breakfast to give him an early start at his work. His children were up as early as their parents, for his eldest daughter, barefooted and with a tin pail in her hand, could be seen going up on the hill among the trees and huckleberry bushes, looking for their cow and followed by a younger sister barefooted like herself.

"Is this your bench, Tom?" Mark inquired of the old fireman whom he had talked to a few hours before.

"Yes, sir, and this water was a mighty help to me," he replied pointing to his pail almost empty resting upon a shelf. The water in this creek tastes so much of the oil from the Lewis's works just above here, and that spring over there tastes still worse, but Mr. Reynolds's spring is so cool and nice.

"Are good springs so scarce about?"

"Just here in this hollow they are, sir, but way up, in that hollow back of Mr. Johnson's house there is another good one, farther however than Mr. Reynolds."

By this time the chargers had eaten their breakfasts, or, eating still of their hot corn bread, were heaping up their charges and praising good humoredly the firemen for having so handsomely during the night disposed of all the coke, but made so many ashes to wheel off. When, after they had shoveled up their piles, they began to clean up and clean their lids, they praised Wag the grinder for having done so much better with his mill the afternoon before. He had left no gravel in his meal, and their caps had not smoked nor blazed during the night.

"Bully for Wag," they all said in unison.

Reynolds heard all this as he walked through them toward his engine where Mark stood observing and listening.

"Good morning, Mr. Mark," said he, opening the oven door of the furnace where Wag was baking his clay.

"Good morning, Mr. Reynolds," answered Mark.

"Now," continued Reynolds addressing Wag sternly, "just keep up a good fire on this grate and be careful to bring up this clay to be as dry and stiff for the mill as you can make it, do you hear? We have had enough of your bad work."

"I could do much better, sir, if my flue was not blocked up. I have told you before that some bricks have fallen in, and, when the stack don't drain well, I can't get a heat as I ought to." He looked sullen as he thus answered and raked his fire, but gave also a rapid glance toward Mark as if he wished to explain his trouble to him.

"Well," replied Reynolds, "you cool off your oven after this batch, and we will have to clean up and repair this flue.

"Glad of it, sir," said Wag.

"It is about five o'clock, Mr. Mark," continued Reynolds, "shall we go up to breakfast?"

"I am ready, sir," answered Mark.

As they went up, they could see Mrs. Reynolds seated inside the door waiting for them. When they sat down at the table, she addressed Mark reverently and said: "Mr. Mark, will you pray for a blessing upon this meal?" He bowed his head in prayer, Reynolds listening silently.

"Did you rest well, Mr. Mark?" she inquired, covering his plate with broiled eggs and handing it to him.

"Very well, I thank you. I did also some writing before I laid down, and, while so doing, I did a thing for which I did not want to wake you up, Mr. Reynolds."

"What was that?"

"Your man old Tom came up and asked whether he could not go to your spring and fill his bucket with some of your water? He said the water from the other spring over the creek and the water in the creek itself are not fit to drink?"

"That's true," said Reynolds's wife. "There ought to be small pipe laid down this hill to carry to a barrel in the ground, in front of the two cabins below, the overflow from our spring." She looked earnestly at her husband as she spoke.

"Well, I'll do that when I come back," said Reynolds.

I'll keep you at your word, thought Mark. Then, addressing Reynolds, he continued: "The water in the creek here and at the works above can be kept clear from oil and tar, by burning both every day under the retorts. Don't your neighbors down the creek to the Elk make complaints for themselves and their cattle?"

"Yes, they do," replied Reynolds, keeping his eyes on his plate, "but I had not thought of this burning for the little tar we throw off. Lewis Brothers above us however ought to do it, for they refine their crude oil as Mr. Fieldbell does with acid and soda."

"How do they reach the Kanawha?"

"They have cut a road of their own through the hills to the licks by Campbell Creek. Their works are not so large as these however."

"And they throw all their tar and acid refuse into the creek?

"They do, sir, and Yankees at that."

"This is an imposition," continued Mark, "which cannot last long. Do they employ Black labor as you do?"

"No, sir. They are three brothers. All work. The eldest is also a physician and carries on a small practice about here, for which he gets good pay in butter, game, and the like. His younger brother keeps the books, as he is not healthy and strong, and the youngest is the refiner. They hire white hands from among the squatters all along the creek and pay them from seventy-five cents to one dollar and a quarter per day, which accounts a good deal for their tar and refuse not being stopped from running into the creek. They are at loggerheads with Mr. Fieldbell about the division line between their land and ours. He says that they are now digging and using coal that belongs to us in that hill over the creek upon which Johnson's house is built."

"He contends that they have passed the line?"

"Yes, by the measurement in the deed of our company from the surveyor's blaze on that old white vale which you see from here midway between the two entries."

"Colonel Albert and Truesdell are of course aware of this dispute?"

"Yes, they are. Johnson first found it out."

"It ought to be settled as soon as possible by mutual agreement and another survey if necessary. Old sores left to run get beyond cure sometimes. How often do you go to town, Mr. Reynolds?"

"Two or three times a month to see Mr. Fieldbell."

"He does not come up here very often?"

"Three or four times, I think, in the last two years."

Had I known this last night, thought Mark, I would have specified it in my letter, but it is sad enough without it.

Upon finishing breakfast, he rose from the table, and, taking from the shelf the bottle of purged oil, he handed it to Reynolds.

"Be careful," said he, "to wrap this up in an old newspaper and to pack it in your bag's saddle pocket so that you take it and bring it back safe and sound, can you?"

"Oh, yes," answered Reynolds.

"I did not write to Mr. Fieldbell, but tell him that I will probably stay with you, since you consent to it, till I hear from Cincinnati within about two weeks."

"I will, sir," answered Reynolds.

"Please also bring along some of the late Cincinnati papers that we may see who is to be our next President in November among so many who are to be nominated."

"Yes, it's pretty hard to tell. Are you a Democrat or a Republican?" inquired Reynolds.

"I am a Republican," answered Mark, "and I believe with Andrew Jackson that the Union must stand and shall be maintained."

"Well, you want *The Cincinnati Commercial*. That's my paper too, and Mr. Fieldbell and Johnson's over there. We are all Republicans on this creek, but the Court House and the Licks are two vile nests of Democrats, for their darkies' sake!"

"It is about charging time," said Mark, looking down toward the works. "I want to be there again. Good-bye till this evening."

"Good morning, sir," answered Reynolds, wrapping up the bottle carefully to tie it up and pack it in his saddle bags. While preparing to clear the table, Reynolds's wife had been listening to the conversation. When Mark had gone, she looked at her husband and said: "I hope, Ed, that you are not going to say anything wrong of this gentleman to Mr. Fieldbell?"

"Well, no, but I want to show him this oil and see what he is going to do about it. It's evidently a big good thing, and the more I think about it, the plainer the way gets to be before me for our return to Cincinnati."

"I will be glad of it if it comes to that. I want to go back to mother, and you would gain by it yourself to return to your machine shop. Mr. Baldwin would take you back at once, for you learned your trade with his father."

"Maybe so," replied Reynolds, as he proceeded to dress himself. He soon went down to the stable, carrying his saddle bags over his shoulder, saddled his horse, and while the charging of the retorts was going on, he started on his ride to the Court House. He had told the teamsters that Johnson would see to their loads for the landing. He did not go to it however on his way to the Kanawha, but rather turned off on a shorter cut among laurel hills.

I don't know but what she is right about it, he thought. I can't make much headway here, and it's more pleasant to work in a live place like Cincinnati than among these dull hills. And if this man really wants to come and work up this business, he can easily beat us both out of it, Fieldbell and myself. I am curious to see how Fieldbell will look when I hand him this bottle, with all his high notions about himself and what he can do.

With these thoughts filling his mind, he rode on until he reached the Elk, two miles above the Court House. The Elk ran there between two deep banks at the foot of high hills, and the top of these banks on each side was just wide enough to furnish room for the track of a wagon, which, in order to let another pass from the contrary direction, had to be led up, pulled by the horses or oxen, high enough upon the hillside, almost at the risk of a general upset, while the wagon which remained in and followed the track, had to be skillfully handled not to be overturned also into the river below. In the winter, when the ground was covered with ice, this narrow road was very dangerous not only for teams, but for riders as well. Reynolds knew this well from personal experience, for many a time he had dismounted at this spot to lead his horse along the two-mile stretch and save it and himself from a fall of fifty feet through the laurel roots and bushes growing out of the ledge of the rocks into the floes of the deep river below.

"What brings you today, Reynolds?" said Fieldbell, surprised, when they both met at the refinery.

"I came to bring you this sample of crude oil with the tar off."

"Why, this is not the bottle which I gave him!" exclaimed Fieldbell.

Reynolds looked at him in surprise also.

"You know then who treated this oil? What bottle do you mean?"

"I gave to Mr. Mark a sample of crude oil here on last Friday, which he must have taken along on Saturday to Mill Creek. Is this it in this new bottle?"

"No, sir. I gave him yesterday myself this sample in this bottle from one of the lower tanks, and two hours after he had it the way it looks now, one third off, all the tar out."

"How did he do that?" inquired Fieldbell with a puzzled look while he turned the bottle in his hands and held it up to admire the clear green color of the oil. He also pulled off the cork to find that the pungent odor of the crude oil had completely disappeared.

"With soda or alkali," replied Reynolds, "but no still."

"No still? How did the alkali look?"

"I did not see it. I was down at the works while he remained at our house after dinner and did this. I asked him at supper whether he would let me bring it to you, as I was coming down."

"What did he say to that?"

"He said yes, made no objection."

He had his reactives in his trunk, thought Fieldbell, so he told me before I sent Joe to Mr. Gork to have him stop this thing. I suspected some test of this kind.

"What do you think of this process, Mr. Fieldbell?"

"I cannot tell anything about it, till I know more."

"It seems to me very simple," added Reynolds, "and the oil he says, can be made to burn though it would clog the wick in two or three hours."

"I suppose it would," grimly answered Fieldbell.

"He said also that the tar could thus all be thrown off at the works, and the cost of hauling could be reduced by that much. Have you seen much of him down here?"

"Within two weeks," he said, "after he has heard from Cincinnati. He is a religious man too after the manner of my wife. He stayed over Sunday with Druker, came to hear preaching with him, and Druker, for a wonder, had his load of oil on board his boat by Saturday night, and did not start down with it to come here till Monday morning at one o'clock by moonlight. He thought also with the preacher that if we stopped charging the retorts on Sunday, we could make more oil by charging only from Monday morning till Saturday night. But you ought to see how he goes round prying into everything about the works. His notions too are right as well, I must say."

"How does he seem to like the place?"

"Well, he is always in good humor, and has nothing to say against it. I must say too I have taken quite a fancy to him."

"What do you think of him?"

"All you say is very much in his favor."

"Does he want this bottle back?"

"Yes, he told me so."

Fieldbell went into his office, emptied the oil into a glass test tube, and tried the gravity of it with a hydrometer, Reynolds watching all his movements.

"This is indeed a fine process," said Fieldbell, drawing up the instrument and wiping it. He then returned the oil into the bottle, and, handing it back to Reynolds, he added, "Reynolds, if Mr. Mark should consent to take your place at Mill Creek, and Mr. Baldwin at Cincinnati would be willing to take you back, would you be willing too? Mr. Mark would evidently be quite an acquisition to our company. I only put the question to you, for I may hear about it from Cincinnati."

"I must speak to my wife first, Mr. Fieldbell, although I think she already smells the rat and would not be against this thing. She wants to be near her mother again."

"Well, talk to her about it, and by and by you let me know."

Reynolds wrapped up the bottle again as he had brought it, and replaced it in his saddle bags, his horse standing at the door of the office.

"Have you the latest *Cincinnati Commercial*, Mr. Fieldbell? Mr. Mark requested me to ask you. He is a Republican too, just as we are."

"Yes, take these along. I am glad he is on the right side with us. Druker is to be here tomorrow with his load?"

"Yes, sir."

"I will want him to boat around to the upper landing a large lot of burning oil for Cincinnati."

"Before I go back, please give me an order on Davis for $5 worth of dry goods which my wife wants," said Reynolds.

Fieldbell wrote it and handed it to him. Both rider and horse were in a few moments out of sight on their way to the store, and thence back to Mill Creek.

At the same time Mark was walking to the landing on the Elk having in his pocket his two letters for home and for Cincinnati, and the bottle which Fieldbell had given him. He had separated the tar from the crude oil as he had done on the other sample which he had given to Reynolds for Fieldbell. He arrived at the landing as Druker and his men, having loaded the boat, were preparing to leave.

They all gladly shook hands with him, Druker promising to mail the two letters and to hand the packed bottle to the steamboat clerk with the understanding that he would call, after the return of the boat, for the receipt from Cincinnati. Mark however said nothing more to him about the contents of the bottle, than that it held crude oil and should be carefully carried, it being also unnecessary that Mr. Fieldbell should know anything about it. It was a private matter.

"Yes," said Druker. "They must handle it carefully, for if they break it, they'll be sorry for the smell of it all over the boat." Mark smiled.

"Your wife and the children all well, Mr. Druker?"

"All well, I thank you. You must stop and see her before you head back."

"I will."

"How do you find things at Mill Creek, Mr. Mark?"

"Pretty much as I expected, not beyond cure however."

"Well, come down again and see us as soon as you can, Mr. Mark."

"I will." And bidding them a safe ride down, he went up to Druker's house. Mrs. Druker insisted that he should later have some lunch before his return to Mill Creek, and while she was preparing it, he went to the schoolhouse, where, standing below one of the windows, he could hear Clark patiently and kindly correcting the spelling or the reading of his scholars, boys and girls, as the turn of each by name was called. Being unwilling to disturb them in the midst of this good work, he managed to return unobserved and quietly to Mrs. Druker's who had fried a few eggs as she had done for her husband a short while before, and, after a pleasant conversation with her as he was partaking of this simple meal seasoned with her fresh butter, fresh milk, and hot corn bread, he took leave again, requesting to explain to Clark the reason of his not calling in, and started back up the creek.

It was past noon when he reached the works. Mrs. Reynolds had also prepared dinner, and she was waiting for him. Telling her how well her friend Mrs. Druker had entertained him, he good humoredly sat down at this second table and helped her do honor to her spread with what renewed appetite he had gained by crossing the many bends of the creek on the logs and gunwales.

Fieldbell mused a long while after Reynolds had left him. I must steal a march on him at Cincinnati, he thought with Mark on his mind, before he communicates to them the advantages of his process. I do wonder how strong his alkali must be, and how much he uses to obtain such a clear riddance of the tar? Might not the refuse of the acid which we use after distillation have strength enough left in it to take down that tar, and how much of it? He tried it at once and repeated the trial several times to find that he could thus purge off one fifth of the tar, but no more. So he wrote to Col. Albert by the same mail which carried Mark's letter brought by Druker that same afternoon but he made no allusion to the sample which he had just received, nor of course to the hint which it had suggested to him for his trial of the waste acid. The deodorization also which he thus obtained was very imperfect, while Mark's was complete. He could however and did claim a reduced cost in the transportation of the crude oil from the mine to the refinery were his process adopted, and he recommended that it should be done at once with Mark at Mill Creek and Reynolds back to Cincinnati. By the boat also which carried Mark's sample, he sent another sample prepared by his acid process packed in a small box.

Col. Albert and Truesdell held a long consultation and waited two days for the receipt by boat of the two samples. The contents of both convinced them that Shortmoore of Newark had indeed sent the very man they needed. Col. Albert answered at once to Mark their approval and acceptance of his proposal, that he, Col. Albert, should come himself to the Court House to take personal supervision of all matters there while Mark would be given the entire management of the crude oil works, and Reynolds be found his old place at the iron works in Cincinnati. He wrote also that he would immediately go to Frankfort and so arrange his affairs that his presence would not be needed there, and that his wife and his eldest son would accompany him to the Kanawha. His son would keep up constant communication as a sort of courier between the refinery and Mill Creek. The question with them was whether he, Mark, would be willing, for the first six months of the trial, to receive one half of the salary which they would pay him when they were satisfied that his process would yield all the advantages which he represented over their present mode of working. This annual salary to be three thousand dollars which they now paid to their representative Fieldbell. He closed by saying that he would wait for his reply before leaving Cincinnati for the Kanawha.

The circumstances under which they had, during the craze of the two preceding years, engaged the services of Fieldbell and agreed to his demand of this high salary were very different from those under which they were now negotiating with Mark. Coal oil was then commanding a much higher price. Few were the men who would be trusted to erect and take charge of such works, and Fieldbell had occupied a pleasant and remunerative position as teacher in a high school at Columbus, Ohio. It was not so with Mark: he was a poor man and in need and search of work, and although Fieldbell had disappointed them greatly in the expectations which he had led them to entertain of his abilities, they had also been taught by this experience to guard themselves against a similar disappointment as far as Mark was concerned. They wished to try him first thoroughly before transferring to him all the authority which they had conferred on Fieldbell. The duplicity of the latter in writing to them the letter which they had received with Mark's had lowered him sadly in their estimation of his moral character.

Col. Albert mailed his own letter to Mark to the care of Fieldbell to whom he wrote also to forward it at once, and that he would probably in a short time take a trip to the Kanawha when they would talk over the acid process as detailed in his letter. In the meantime, the result of his consultation with Col. Truesdell was that he had proposed to Mark to take charge in certain conditions of the crude oil works upon an order to Reynolds to this effect enclosed in his letter.

When both letters reached Fieldbell, after reading his own, he looked anxiously at the other for Mark and turned it over several times severely tempted to open and read it and then seal it again. But the fear of being detected kept his hand off, and he sent it by Druker to hand it to one of the teamsters at the landing. By the same mail he received another post marked Newark, Ohio, the handwriting of the address also to his care showing it to be that of a lady, doubtless his wife, thought Fieldbell.

It was eleven o'clock when the teamster handed both letters to Mark who stood then near the engine conversing with Reynolds. He went and sat on a pile of bricks in the rear of the retorts where he would not be disturbed and read first his wife's letter. It was with difficulty that he could read it at first, for warm tears blinded his sight and ran down his face till at last he mastered his emotion and began, more collected, a second reading of it.

"My dear husband,

You may well conceive the joy with which we received each one of your three letters from Cincinnati, from the Kanawha Court House, and the last from Mill Creek. In this last you say that I can now write to you to the care of Mr. Fieldbell, the President there of the company whose other officers you describe to me in your second letter from Cincinnati. This Mr. Albert seems to be a good man, just as Mr. Shortmoore told you that he was, and his other representations of the present condition of the company seem to tally well with what you have found their affairs to be on the Kanawha. They are indeed in bad hands, and it seems to me very clear that those Kentucky gentlemen will be glad, as soon as they get well acquainted with you, to secure your service. As to the difficulties which you anticipate for training up our children in that wild country, would they not urge us to increased vigilance and tenderness over them, especially should the leadings of Providence appear very plain to us for our acceptance of a call to go there?

"Sure it would be that our trust in God's promises, while obeying this call, would be its own reward, especially also if we do not consult our own ease, but with the work, and also the good of those poor people among whom this work would thus be cast. All your friends here are very kind to us. The men whom you trained and left carry on the alkali works to the satisfaction of Mr. McCune and of Mr. Caze. John, after his day's work, comes and plants in the garden all the seeds I give him. He is a clear worker, quick and intelligent. Samuel and Harriet help him too with all their might, and Nathan, not being able to walk yet, looks at them all the time, seated upon the doorstep under the porch. They will soon have another little brother or sister, and then they will be four lives together one after another to make life dearer to us both for the good seed which God will give us every day to plant into their little hearts, there to be watched over and watered by us that He may be pleased to give it the promised increase in His own time and in His own way to the praise of the glory of His grace, in their usefulness as long as they shall live and in their salvation at last. We all kiss you the kiss of love which you have so well taught us.

Write soon again –
Sarah."

After folding this letter and enclosing it again in its envelope, he slipped it carefully into his coat pocket, and opened the other from Col. Albert. As he read their manly acceptance of his proposal, followed by their counterproposal of a remuneration amply sufficient for the moderate wants of his family, also the Colonel's purpose to come up on his own, the conviction filled his mind that his wife's letter had been brought providentially with the Colonel's, to counsel him to say yes to it all, in which case he was to read another letter enclosed, addressed to Reynolds and to deliver it to him as an open one, if he objected to none of its contents. It read thus:

"Mr. Ed Reynolds. Dear Sir. We propose by this mail to Mr. Mark, who is now with you, to take charge of all our works at Mill Creek. We need his practical knowledge and his experience to make these works the success which we desire. Mr. Baldwin here is perfectly willing to take you back to the same work and at the same wages which you had with him two years ago. We will pay the expenses of your return and your wife's, including the freight on your goods, also your bill against Mr. Mark, approved by him, for what furniture, garden truck, and other things he might wish to purchase up shortly from you. As I propose going to the Kanawha, have your account with Mr. Fieldbell ready for settlement. We write to him about it. We send this to Mr. Mark that he may read it and then hand it to you, if he will. Truly yours Col. Albert."

After reading this third letter, Mark arose from his seat on the bricks, and saw Reynolds at his engine looking intently toward him.

"Mr. Reynolds," said he, when he had approached him near enough to speak low, not to be overheard by the men, "I have an open letter for you from Col. Albert which he has instructed me to read before handing it to you."

"What is it about?" answered Reynolds taking and unfolding it to read. As he did so slowly, his countenance fell and he knit his brows.

"Short notice this," he said, looking again at Mark and folding the letter. "Short notice, but it's all right, if you accept their offer, Mr. Mark. Do you?"

"How do you like yourself the idea of living again in Cincinnati, Mr. Reynolds?"

"Oh! 'Tis all the same to me. My wife will jump at it however to be near her mother once more. If you accept, I would like to sell you Selim, my horse, a splendid one. I have had him these two years and a better, gentler beast you never saw under the saddle or in harness. Our old cow also I would like to sell you. You know what good butter she makes, and, with her, our sow and her litter of pigs, the chickens too, and all that spring truck up there around our house. But it is twelve o'clock, so let us go to dinner."

He blew the whistle for all hands in the works and in the bank to go to dinner, and, followed by Mark, he went up the path to his house.

"You have not said yet whether you accept or not, Mr. Mark?"

"I think I will under the circumstances and since the change does not appear to work hard on yourself and would rather be pleasant to your wife, you say."

"Well, I don't know but what you are right. These gentlemen are clever men and do things right with all who work for them. I don't blame them for trying to make more money out of these retorts. Everything here has cost dear enough, and I take it that you are the very man to handle them."

Mrs. Reynolds, from the house, had seen both her husband and Mark reading letters, and talking together with no little concern. She had set the table, and when they came in, she looked at Mark and said: "Have you heard from your wife and from Cincinnati, Mr. Mark?"

"Yes, Mrs. Reynolds. She and the children are all well. Your husband has also some news for you, which, he says, will please you."

"I know what it is," she exclaimed joyfully clapping her hands. "We are going back home to mother."

"Well, that's about it, Jane," said Reynolds, handing to her the letter of Col. Albert. She was not long reading it. Her face beamed with gladness, and her eyes filled with tears which she could not repress.

"Let us sit down, Mr. Mark," and, with voice trembling with emotion, she added: "Please ask for a blessing upon this meal." Mark offered thanks for all the mercies of the day, for the food before them, their continued health, strength, and reason, and especially prayed for a submissive and contented spirit under the news which they had just received.

"Do you accept, Mr. Mark?" she inquired eagerly after filling his plate.

"I have just told to Mr. Reynolds that I thought I would under the circumstances."

"Well 'tis of the Lord," she added. "For all our times are in His hands and He makes all things to work together for our good."

"The good of all who love Him," he rejoined, "for the love to us is the love of perfect wisdom, and He knows best which is best for us" [Romans 8:28].

"I believe that Mr. Mark, good for nothing fellow that I am," added Reynolds. "What will you give me for Selim?"

"Col. Albert does not name him, does he among the things for your bill against me? Look on his letter."

" 'Tis true," replied Reynolds glancing at it. "He writes of furniture, garden truck, and other things. You might not wish to purchase the horse from me, but you will need a horse to go to town, and a cow for butter and milk."

"I shall have to wait, Mr. Reynolds, till I see my way clear for this purchase, much as I like both Selim and Muly."

"Oh! We can make an arrangement for that," said Mrs. Reynolds. "I want your wife to have my good cow. She is so gentle, and I know she will take to your children."

"I will propose the thing to Col. Albert when we meet him on the boat which will bring him up," said Reynolds, "for I want also old Selim in good hands after I leave him."

In this spirit the conversation went on, when, hearing the steps of someone at the door which was opened, they turned and saw Fieldbell standing with a dubious smile on his face.

"How do you do, Mr. Mark?" said he.

"Very well, I thank you," replied Mark advancing toward him. "I am glad to see you."

"We have not seen your face for so long," said Mrs. Reynolds, "that I had almost forgotten how you look, Mr. Fieldbell, till Mr. Mark named you."

"Sit down and take dinner with us," added Reynolds, "poor as it is. Where is your horse?"

"I left it below with Bob, till I come back, and told him to give him a few ears of corn. I will sit down and help you to finish your dinner. Mr. Mark, have you the two letters which I sent you this morning?"

"I have, sir. One is from Col. Albert who enclosed another for Mr. Reynolds, which I have handed to him."

"Yes, and I was talking a bill of sale of all our things here to Mark," said Reynolds.

"So you accept then, Mr. Mark?"

"I cannot well do anything else," replied Mark. "Col. Albert will be here in a short time to complete the arrangement. You have this understanding from him, have you not?"

"Yes," replied Fieldbell looking inquiringly at Reynolds and his wife, as if he wanted them to explain.

"We are going back to Cincinnati, Mr. Fieldbell," said Mrs. Reynolds, "and glad I am soon to meet my mother again and remain with her."

"The Colonel," said Fieldbell, "wrote to me that he had proposed to Mark to take charge of these works, and that he would be up here shortly. But he did not explain himself as fully as he said he would when here. Hence this trip of mine."

"Ed, show your letter to Mr. Fieldbell," said his wife to Reynolds.

"Here it is, Mr. Fieldbell," added Reynolds. Fieldbell took and read it attentively.

"Mr. Mark," said he, returning it to Reynolds. "Your handing this to Reynolds shows of course your acceptance of the Colonel's proposal?"

"Yes, sir, it does, and I am glad to see that it has brought up no bitterness between us. Mr. and Mrs. Reynolds have a home in Cincinnati where they will find a joyful welcome."

"God be thanked, we will," exclaimed Mrs. Reynolds. "Not that I am discontented here, but home is home."

"Just so," added Reynolds.

"Well," added Fieldbell, "what begins well, ends well. Is this the sample of the crude oil which you sent by Reynolds, Mr. Mark?" he continued, taking hold of the bottle on the shelf.

"Yes, sir," replied Mark.

Fieldbell pulled off the cork and smelled the contents.

"No odor about it," said he. "How strong is your alkali for this, Mr. Mark?"

Mark smiled and answered: "Mr. Fieldbell, I worked hard and many months to find this. We will talk about it after the Colonel gets here."

Reynolds gave a significant look at his wife, and Fieldbell continued turning the bottle in his hands, unwilling to manifest any disappointment, especially since he could express none by word of mouth. Root, hog, or die, thought Reynolds of him.

Dinner was over by this time. Turning to go down to the works with Mark and Reynolds, Fieldbell shook hands with Mrs. Reynolds and bade her good-bye.

"When shall I expect you down, Reynolds?" said he, as he went on the steps after him.

"As soon as you send me word that the Colonel has come up. I'll then make up a wagon load of what things we will take to Cincinnati and we will come down the next day."

"In the meantime, settle up with all the hands and with Johnson to this day, and send the accounts to me before you come down, that I may be ready to pay them all any time they wish. Mr. Mark wants to take charge of this now in his own way."

"I will do so," replied Reynolds.

Chapter Six

Radical Improvements

As they went among the retorts and the tanks, a wondering look appeared upon every countenance of the firemen and other hands. The wheelers also took the word into the bank. The deferential manner of Fieldbell toward Mark seemed to please the men, and the clean appearance of the floors below the fire fronts evidenced the care with which all the fires were kept up.

"It is not far from two o'clock," said Fieldbell, "time for me to return. Johnson is busy in the bank, I suppose, for I did not see him."

"Yes, sir," answered Reynolds. "The mine has been doing a good deal of porting lately." Fieldbell's horse was tied at the fence near the Black hands' quarters, and one of them, after loosening the bit, had thrown upon the ground before him a few ears of corn which he was still munching and grinding.

"Shall I write to Col. Albert that you have accepted his proposal, Mr. Mark?"

"Yes, sir, you may do so, and please tell him that I will write also today or tomorrow. Someone passing down this creek to the Kanawha will take my letter to the post office."

"If he mails an answer to your care before he comes, please send it to me at once."

"I will do so, Mr. Mark." Fieldbell then mounted his horse, and, bowing to them, was soon on his way to the Elk, not much wiser than before about Mark's process, but satisfied that this Mark was no common man to have tamed so completely, so quietly, and in so short a time a crook like Reynolds.

What will be the outcome of all this? he thought as he rode along. When, after the Colonel has been here but a while, he will learn all about this process, and will want it carried out. What of me then? Could I not represent to him that my acid process would utilize our spent acid which we now throw off, and would thus cost so much less than that alkali process which would necessitate the erection at Mill Creek of an apparatus to manufacture it? And as to the odor still remaining after the use of my process, the distillation would remove it. The crude oil purged by Mark's alkali would have also to be distilled anyhow. The Colonel is but an old printer. He don't know these technical points, but he cares much for not spending any more money in putting up new fixtures. Even Mark's neat purging would reduce the quantity of acid now used and turn out a better burning oil commanding a higher price than either the oil we now make or that from my acid process. I must paddle my boat carefully through all these breakers, for this letter of the Colonel to me is very short and reticent, and between the lines I can read a great deal in favor of this Mark.

After his supper at the hotel, he wrote to Col. Albert that he had received his letter, had forwarded the two for Mark, and had just returned from Mill Creek, where he had found Mark and Reynolds both willing to the proposed change, and that Mark intended to reply also by the next mail. Suspecting, from the expressed readiness of the Colonel to come up the Kanawha, that Mark might have disclosed to them the evident advantages of his alkali process, he laid much stress upon the economic advantages of his own, as he had ruminated them during his return ride. Possession is nine points of the law, thought he again to himself. I am President here of this company. They are all well off. I have a good salary and a pleasant position. I will certainly do all I can to keep it, even if I have to lie not a little to do so.

He mailed his letter the same evening. In the afternoon of the next day, one of the squatters on the headwaters of Mill Creek mailed also two letters from Mark, one for his wife and the other his reply accepting the counterproposal of the Colonel and stating his friendly way in that Reynolds, through his good wife's influence over him, had accepted the change. He also related the unexpected visit of Fieldbell and how pleased he seemed to be while handling and smelling the second sample of purged oil which Reynolds had already carried to him and brought back, a fact which Col. Albert was not aware of when he had written to Fieldbell his receipt of his own sample to consult with him about it during his intended visit to the Kanawha.

Col. Truesdell was the first one who noticed that both letters arrived at the same time. He laughingly handed them to his older friend Col. Albert, telling him to commence with Fieldbell's letter and then to go on with Mark's.

"Well, well," exclaimed the old Colonel laughing also as he terminated his reading. "Has not Fieldbell got himself into a nice mouse trap with all his finesse. However, there is doubtless some truth about the economy of which he speaks in relation to his process, his process—yes, *his* process, the scamp! And why had he not thought of it these two years past without waiting for Mark to put him on the trail! Truth will even out. I will see into all that when I get there."

"When will you go?"

"By the *Ellen Grey* this very week on my return from Frankfort with my wife and George."

"You had better write at once then to both Fieldbell and Mark, so that they may be on the lookout for you. Tell Mark to wait for your coming to see him at Mill Creek. Reynolds will then be out of the way too, and Mark will feel no embarrassment in his talk with you."

"Yes, I will write to them immediately and also to friend Lorme of the bank for two rooms in his large house. Boarding with his pleasant family will suit us exactly. My wife's health is not very good, and this change of air and habits, I think, will do her good."

"Yes, she will like to live with Mrs. Lorme and her old mother, Mrs. Dunn."

"And she will not miss your Frankfort church, for Dr. Brown, who preaches in the Presbyterian Church at Kanawha, is counted one of the most eloquent preachers in Virginia."

"She has heard of him, and will be glad to hear him. There are some nice families too, just as much as in Frankfort, and I think she will soon have friends who will try their best to make the place pleasant to her."

The three letters were written and mailed that morning, and, two days after, the Colonel, his wife, and their eldest son George took passage on the *Ellen Grey* just as the stirring news had arrived of the nomination by the Charleston Democratic Convention, on the 23d of this month of April 1860, of John C. Breckinridge of Kentucky and of J. C. Lane of Oregon, for President and Vice President of the United States.

The Colonel had been a lifelong friend of Henry Clay, and was now a pronounced Southern Union man, and, with all loyal Kentuckians, felt very anxious about the other two nominations to be made on the 16th and the 19th of May following, at Chicago and at Baltimore, the first by the Republican National Convention, and the second by the Constitutional Union Convention. His sympathies were centered upon the latter. The heated, angry discussions and scenes which had ended in the final disruption of the Charleston Convention had cast before all minds fearful forebodings of a great evil approaching, and during the passage up the Ohio and the Kanawha, the excitement in his mind found relief in animated conversations with his fellow travelers, most of whom he found as earnestly attached as himself to the Constitution and to the Union. Not a few however declared themselves in favor of the Republican National Convention. He argued with them also, but could not admit the correctness of their radical demands for the extinguishment of slavery, unless it was to be gradual and scaled on a pecuniary basis of compensation. Fewer still he found boldly advocating the arrogant claims of the Southern leaders for its acknowledgement in all the States and Territories. With them he would not waste any converse. He thought them incurable lunatics and passed them by with a courteous bow. Should the men, he thought, succeed in bringing about the dreaded armed conflict which they so loudly and so persistently threatened if their claims were not conceded, what is to become of his investment and of his friends on the Kanawha and of their possessions in Kentucky as well? How ought he to allow himself to remain away from his native state, which would soon perhaps need the help of all her loyal sons to keep her in the Union?

These anxious thoughts engrossed much of his mind when the boat arrived at the mouth of the Elk. From there he pointed to his wife and to his son the tall frame structure of the refinery looming up high on the bank of the tributary. A few moments after, they reached the upper landing. In the small crowd which awaited the casting of the stage planks, he descried Fieldbell and Lorme who saluted him and soon pressed their way up to the cabins where the Colonel introduced both to his wife and to his son George.

"You have received my letter, Mr. Lorme?"

"Yes, Colonel, and we will be glad to have you all three board with us, that is if you like the two rooms and our Kanawha fare."

"I thank you. We are old-fashioned folks, not hard to please. Did you forward my letter to Mr. Mark, Mr. Fieldbell?"

"I did, sir. I will now send word to Reynolds to come down with his wife as I promised him, now you are here."

"Very well. Now we will go up with you, Mr. Lorme. George, you see to the baggage."

"The dray is down there on the wharf waiting for it, Colonel," said Fieldbell, "and I will remain here with your son to see him and the load all safe to Mr. Lorme's."

Guided by the latter, the Colonel and his wife went up to his house, and were cordially received by Mrs. Lorme, a New England lady, and by her aged mother whose benevolent looks and ways won at first sight the heart of Mrs. Albert. Their ages were also about equal, and if her daughter was more demonstrative in speech and manner, still it was evident that she had received a careful New England training. Mr. Lorme was a Virginian gentleman, cool and quiet, a teller in the bank. The two rooms above were visited, surveyed, and found pleasant in every aspect, the front one facing the street and the other opening upon a veranda which overlooked a large garden in the rear of the house.

Soon after, the baggage arrived, was brought up, and, after dinner, the rest of the afternoon was spent in opening the trunks and finding a place for everything in the closets and the bureaux.

So he has come, not on a trip, but to stay, thought Fieldbell. Indeed, the information had dawned upon him from the moment Mr. Lorme had, before the arrival of the boat, communicated to him the inquiry of the Colonel about board and two rooms for himself, his wife, and his son.

Well, Fieldbell continued, this is a good move. It will work well. Mark is a clever man. He will keep everything straight at Mill Creek, and, on the score of economy, I think I will succeed through the Colonel in staving off Mark's purging process and in having mine adopted. I will try hard anyhow.

After breakfast, the next morning, the Colonel with his son walked down to the refinery and found Fieldbell in his overalls, busily engaged among his men, some of whom were firing and watching the stills, while others, in the high platforms, were keeping the large iron agitators in rapid motion. Still others were filling bags and piling and pinioning them singly between the square boards, and piling them upon powerful presses, which, when thus made ready, they would subject to a gradual, heavy leverage to extract and separate all the oil from the paraffin, reduced thus to as dry a state as possible. The oil was then packed in barrels.

The Colonel, aided by Fieldbell, spent several hours examining carefully the various stages of this refining work. He could see that it demanded the undivided time and attention of an experienced, intelligent manager not only during the day, but often also during the night, and that as long as ten miles stood between this refinery and the crude oil works, so long that it was necessary that there should be two different managements, both however working together in harmony under two intelligent heads.

This I must bring about, the Colonel thought, for peace's and our company's sakes. At the same time, I must have an honest talk with this man and have him understand that he cannot deceive us. He was standing then in the rear of one of the high crude oil stills and was looking at the pile of pitch which a man from the top, outside, was throwing down by bucketfuls as they were handed up to him by another man, breaking it up from the bottom inside the hot still.

"Fieldbell," said he after musing a while. "Your process would not purge all this pitch out of the crude oil as neatly as Mr. Mark's, would it?" Fieldbell's face reddened slightly from surprise and confusion, but, recovering promptly his self-possession, he replied: "No, sir, but have you learned from Mr. Mark the cost of his fixtures to prepare his alkali and the cost of the alkali itself?"

"No, not yet."

"On the other hand, the acid which I propose to use is a waste with us. We throw it away."

"That's true but would not his process lessen considerably the quantity of acid which you would use under your process?"

"I cannot tell. It might and might not."

"Would not his process also reduce your cost for distillation even under the process you propose?"

"That's another question."

"It would evidently reduce greatly your present cost on the work of all these stills?"

"Yes."

"And with all the tar out of the way, in the first place, as he proposes, would not your distillation yield a very fine burning oil?"

"Yes, it would, but what of the cost of the alkali and fixtures, as I said before?"

"Well, we must look at that very carefully of course after Reynolds and his wife have come down. I will go up and consult with Mr. Mark about this cost."

But, thought the Colonel to himself, it could not be so very high in Newark, since Shortmoore has proposed to Mr. Mark to purchase the plants.

This frank and fair way of broaching the question satisfied Fieldbell of the honest intentions of Col. Albert, and convinced him very timely that Mark's correspondence and talks had evidently shed a deal of technical light in the mind of the Colonel upon important technical questions pertaining to their craft. I can't lead him off so easily, he thought, the old teacher.

"Have you Reynolds's settlements with all the hands in Mill Creek?" inquired the Colonel.

"Yes, sir, he sent them to me. I have them all in the office."

"Let us go there now. I will relieve you of all this office work, so that your mind may be more free, and you may have more time to attend to all this refining. I see it's a big job, and, mark my word, it's going to grow."

"It will indeed be quite a relief," answered Fieldbell. "I am very glad you have come to stay. Between you and Mr. Mark, I will be able to do more work at refining and more satisfactory too."

"I think so decidedly. How soon do you think Reynolds will be here tomorrow?"

"He will probably make an early start in order to see you tomorrow and be ready, with his wife to take passage at night on the *Ellen Frey*."

The office stood off at a short distance from the refinery in the corner of a large clover field which belonged to the company, and in the center of which had been dug a large deep well. Into this well all the waste tar and spent acids were run off, through a large underground trough, from the refinery to keep this refuse from going into the Elk River. It was a moderately large frame room lighted by two windows filled with strong shutters. Its furniture mainly consisted of a large desk, a long table, a few chairs, an iron safe, and a bed. Tools and factory supplies occupied some shelves in the closet. In it, Fieldbell changed his clothes, and sometimes slept during dark and stormy nights.

He showed all his books and accounts to the Colonel, and, in the midst of his explanations, was soon called off by one of the men for some personal supervision at one of the agitators. Col. Albert remained with his son pursuing his examination of all the contents of the desk and of the safe, and when Fieldbell returned at dinner time the whistle having just been blown, he asked him whether, in order that they might see still more of each other, he would not be willing to board also at Mr. Lorme's, if the latter would accommodate him.

"I will be glad to do so, Colonel, if you will arrange it with him."

"Well, come with me. We will dine together." Fieldbell was well acquainted with Mr. Lorme, whose wife and mother knew him from his two years' residence in the place, and it was soon arranged that he should also become one of the family and have his baggage brought from the hotel to his new room at Mr. Lorme's. Col. Albert did not go down to the refinery after dinner and spent the afternoon renewing his acquaintance with the leading citizens of the Court House whom he had met once a few years before on the occasion of his first visit with his Kentucky friends when the refinery was being erected, the mine opened, the road cleared along the creek, and the crude oil works were being built up to receive the retorts. For his genial spirit was generally known and he was very hearty. He could tell a good story, and he was soon an outspoken favorite with all parties, though an outspoken and resolute Constitutional Union man.

In the evening, Dr. Brown and his wife, an aged couple, who had resided for many years near the Court House, called a pastoral welcome to the Colonel and his wife, and the pleasant hour, thus spent, formed the first link of a golden chain of sacred associations, which, under his faithful pulpit ministrations, bound Colonel Albert and the two families close together for months, until the rude shocks of fratricidal war, like a storm, broke it apart, and divided them hither and thither. The venerable Virginia preacher, still clinging to his desolated home, died of a broken heart, while the Colonel, in Kentucky, became the trusted friend and adviser of Abraham Lincoln for whom he had not voted.

The next day, late in the forenoon, the Colonel was seated in the office of the refinery, writing a long letter to Col. Truesdell, when he turned at the sound of some footsteps before the open door, and saw Reynolds standing and looking at him.

"Col. Albert, sir?"

"Yes, you are Reynolds, if I remember right? Where is your good wife?"

"There she is, sir, in that wagon on the riverbank, back of the refinery, waiting for me. Mr. Fieldbell just now told me that you was here and would have come with me, but could not well leave the work he was at. He will soon be here however."

"Well, tell the driver to bring your wagon this way, Reynolds. I want to see your wife also. You must have started very early this morning?"

"We did, sir, at five o'clock. It is not so bad for ten miles of such a road with a pretty good load of what things we kept, and a stoppage for good-bye to Mrs. Druker and her husband at their house on the Elk, but I will go and tell John, who drives the wagon, to come down."

Fieldbell soon came from the refinery and was speaking with the Colonel when the wagon stopped at the door. They both rose from their seats and saw Reynolds's wife seated in a chair in front of trunks and boxes. She was dressed in her best and her face was radiant with smiles as she looked at them.

"Mrs. Reynolds," said the Colonel, extending to her his right hand, "I am indeed glad to make your acquaintance, though I knew you well before this as the better half of the good for nothing feller," he added smiling also as he turned toward Reynolds, who took the term in good part, almost, signifying by the expression upon his countenance that he had well earned it.

"Going back to your mother, Mrs. Reynolds?"

"Yes, Colonel, it will be surprise to her also, for she don't know that we are coming. It is two years since I have seen her!"

"Well, it's all right. Mr. Baldwin, with whom I had a long talk in Cincinnati, is on the lookout for Reynolds, and Reynolds will feel more at home making engines than making oil. My wife would like very much to see you before you go on the boat. Reynolds, you know where Mr. Lorme lives, the teller at the bank. We board with him. You drive your wagon to his door and tell my wife to take care of Mrs. Reynolds while you go down to the boat to engage your passage. I will meet you there, at the office of the boat, to take you to dinner with us, after which we will come back here to settle up all our accounts. You have Mr. Mark's order for what things he wanted from you?"

"Yes, sir, here it is on this $82 bill for furniture, garden truck, sow and litter, and chickens, but my horse and our cow are not in it. He said he was not prepared to buy the horse just at present, nor the cow either, till his family come. The horse, he will be too busy, he thinks, up there to need it to get anywhere. I left them both in his care however, and Mrs. Johnson will look after the cow for her milk."

"How much do you want for your horse?"

"One hundred dollars, what I paid for it a year ago. His name is Selim, a fine animal."

"Well, George, my son will try it as he will need one. If it suits, you will get the money for it in Cincinnati. I will send you an order on Col. Truesdell. And how much for the saddle and bridle?"

"Ten dollars more."

"Very well and for the cow, how much?"

"It costs me thirty dollars, and we don't want to sell her for one cent less, do we, Jane?"

"I will pay you thirty dollars for the cow, Mrs. Reynolds." Mrs. Reynolds smiled her assent and added: "She is indeed well worth the money and the name, Colonel."

"Well then, Mrs. Mark, I have no doubt, will be glad to get her at that price when she comes. When I settle with you, Reynolds, after dinner, I will include this $82 in your account against us. Now, go on to Mr. Lorme's house. Leave your wife there with mine and go down to the boat. I will meet you at the office shortly before twelve o'clock. Good-bye, Mrs. Reynolds."

She bowed, answering "good-bye." Reynolds having seated himself in another chair next to hers, the driver started his horses to follow the street which bordered half a mile on the Elk, till its junction with the Kanawha up which they went along the main street, passing the bank. Just beyond it, they stopped before Lorme's large and commodious stone house. Mrs. Albert expected them as per agreement in the morning with her husband, who had also requested Mr. and Mrs. Lorme to let him extend this proof of kind feeling toward Mrs. Reynolds and her husband on their return home. The cordial greeting which they thus received from the entire family added much to the high regard that Mrs. Reynolds had always entertained for the Colonel, although she had before seen but very little of him.

Leaving her there, Reynolds went down with the wagon to the boat, engaged their passage, made choices on a state room, and had all their baggage brought on board. The last box was being carried up from the wagon when the Colonel and Fieldbell made their appearance, and, after a short talk with the captain, they went up to the house.

"How did you leave Mr. Mark, Reynolds?"

"In good spirits, sir. He expects to see you up there very shortly."

"Yes, I think I will go with my son George, perhaps tomorrow morning we will hire some buggy. I hope the road won't be too rough on the buggy wheels and springs?"

"Oh no, but drive carefully over the big stones in crossing the creek."

The dinner passed off pleasantly. Reynolds's ways were somewhat awkward at first, but he was soon set at his ease by the kind questions put to him and to his wife, especially by the three ladies whom the gentle and unaffected manners and words of Mrs. Reynolds had also won over. After dinner, the engineer followed the Colonel and Fieldbell to the office at the refinery. In two hours, he explained all his statements of accounts with the hands in Mill Creek, was paid the balance of salary together with his bill against Mark, and received an order on the cashier in Cincinnati for the fare on the boat for himself and his wife, and the freight bill on their extra baggage. The cow was also paid for.

"Now, Reynolds," said the Colonel, "we part in peace and with good will for each other. The boat will leave in about two hours. I will go back with you to the bank that you may cash this check. Then we will stop at the house for your wife, and I will take you both on board."

Reynolds had been a Sabbath school boy once, and the recollection of his people deserving all this kindness brought up into his mind, contrite for once, a promise which had always appeared strange to him, but which somehow it seemed almost to him that he could realize, as applying however to the Colonel: "In so doing, then shall ye heap coals of fire on his head" [Proverbs 25:22]. But the head is mine, if the coals are the Colonel's, he thought and thought again. And his voice trembled as he shook hands both with the Colonel and Fieldbell.

To the bank they both went. They passed along the street some storekeepers who knew him well, and who had heard of his going away. Reynolds saw him as they stood on their doorsteps and gave him a friendly greeting in the presence of the Colonel whom he introduced to them. The latter was pleased, for he did not want to be a stranger to anyone in his new home, and his genial disposition delighted in such expressions of friendliness. From the bank, where Reynolds took also leave of Lorme, they went to the house. Mrs. Reynolds was in the parlor with the ladies waiting patiently. She thanked them all for their welcome and kindness, and in a few moments, Reynolds and his wife were on board the boat. The captain had already rung the first and the second bells.

"Is Mrs. Johnson as good a wife and woman as Mrs. Druker, Mrs. Reynolds?" inquired the Colonel as he presented her a chair in the ladies cabin.

"She's a Virginian also, Colonel, but Mrs. Druker was raised in east Virginia, in the valley, and had advantages for education which Mrs. Johnson, when a girl, did not get on this river up at Paint Creek, another place like Mill Creek. She was born and raised there among the miners and salt men till she got married, but for all that, she is a good wife and a good mother in her way, hard working too, and kind-hearted. I know you and Mr. Mark will like her. Mr. Mark is taking his meals there now since we left, and she thinks there is nobody like him, except her husband who speaks very highly of him too."

Here the bell rang for the third and last time. The Colonel bade them both a pleasant and safe journey, recommended them again to the kind attentions of the captain and descended the steps. As he stood on the landing, talking with some gentlemen and looking at some last barrels of their burning oil which the crew were rolling up, the stage planks were drawn and wheeled in on the foredeck, and the boat's whistle was blowing for the starting turn of the wheels. Reynolds and his wife stood together on the upper deck before the door of their state room and bowed several times to the Colonel as he went up the landing sending to them with his right hand a final greeting, but thinking within himself: "If he has wronged us badly, we have wronged him also by letting and leaving so many temptations in his way. I would have prevented this, had I come and settled here two years ago, and I would have done it as well as now."

Early the next day the Colonel and his son in an express wagon hired from the hotel, the father holding the reins and George, his loaded rifle straight up between his knees, were winding up their way to Mill Creek. While they were crossing, just after leaving the Elk on the left, the horse gave a sudden snort, pricked up his ears, and wildly threw up his head on one side, his eyes fixed upon the lower branches of one of the many beech trees which lined the road on each side. His trot also turned into a run which the Colonel tried at once to check by pulling hard on the reins.

"What does he see?" he exclaimed to his son.

"That panther or catamount. Whatever it is in that tree!" answered George, quickly raising his rifle and pressing his finger on the trigger. The spring foliage could only partially hide the crouching body of the animal. It was almost as large as a mastiff's but its head and glittering eyes, like those of a tiger, could be seen distinctly.

"Don't fire, George! We have no time to waste on the brute. Don't provoke it, leave it alone, unless it comes after us!" But it did not. They rapidly passed on, and the Colonel soon succeeded in bringing the horse back to a fast trot.

"We will go on to the landing and stop to see Druker and his wife before we turn off on Mill Creek. I want to examine that landing about an improvement there of which Mr. Mark wrote to us."

"What improvement, father?"

"He wants to do away with the barrels altogether and replace them by an iron tank on wheels filled at the works, and to be emptied into the boat through an iron pipe."

"The boat to be made tight first?"

"Yes, and decked over."

"That would be a good way to save time and work."

"It would indeed, and we could sell the barrels. The change would soon pay for itself."

At the landing, they found Druker and his four men busily engaged in loading the boat. The Colonel had not seen him for two years, but soon recognized him by his bearing and his talk to the men.

"Capt. Druker?"

"That's my name, sir, and if I am not mistaken, you must be Col. Albert? Reynolds told me you was coming up."

"Yes, you are right. I am the man, and this is my son, Captain. We are on our way to Mr. Mark at Mill Creek. Is that your house, up there?"

"Yes, Colonel. Come and take each of you a glass of pure milk. My wife will be glad to see you again, although she had but a glimpse of you when you came up two years ago."

"So will I be glad to see her again, and get better acquainted with you both this time." They had both got out of the wagon, and George had fastened the horse to one of the corner ports of the shed. Druker led the way to the house.

"You have a first-rate man, Colonel, at Mill Creek now," said he as they walked along. "A real gentleman too who will work for you there as he would for himself."

"This is what he writes about you, Druker."

"Does he indeed?" exclaimed Druker laughing. "Well, we must feel alike then, and I must keep myself straight to live up to the good opinion of a good man like him." His wife was at the door.

"This is my wife, Colonel. If you recollect her, Sally is her name. Sally, this is Col. Albert and his son. I told them of your fresh milk, and I have brought them to have a taste of it and see you."

"I remember you well, Colonel," she said. "You was here but a few moments about two years ago, you first came with Col. Truesdell and some other gentlemen."

"Yes, you had not been long here yourself, and your husband was working hard bringing up the retorts on the boat. Both your children at the school up there?"

"Yes, sir. We have a fine school and a good teacher for it, Mr. Clark."

"Yes, I have heard of him through Mr. Mark. I think we will have time to stop again and see him when we come back from Mill Creek this afternoon, that is if the school is not out and he has not gone home yet."

"Oh! He does not leave the children out till 4 o'clock, and he will be looking for you, for somehow he knows that you feel an interest also in his work, just as Mr. Mark does."

"I am glad he thinks so. We are indeed all in the same boat. Mr. Mark has informed us of all the good done in this school, and especially of the liberality of the people around here to keep it going almost all the year round."

"Yes, sir," said Druker. "We want our children to get as much education as we can give them. We know too well how little of it we have got ourselves."

"Come in, gentlemen," said Mrs. Druker. "I have a pitcher on the table full of fresh milk for you, these cold biscuits also for a little lunch till you get to Mill Creek."

"We thank you, Mrs. Druker," replied the Colonel coming to the table followed by his son. She filled three large tumblers which she had also brought. The long ride had sharpened their appetite, and they found the milk so rich and refreshing that they gladly drank each another glass full and ate of the biscuits not a few. The Colonel, while talking, looked around and admired the neatness of the room in which they stood. The floor was bare but faultless in its cleanness. It was evidently scrubbed every week. The kinks between the hewn logs were carefully filled with mortar and coated with a whitewash which had of course been extended to the under part of the wide boards that formed the ceiling. Against an opening in the latter, at one end of the room, a step ladder leaned, by which access was gained to the story between the ceiling and the shingled roof. At the other end of the room, the center of which was occupied by the hearth, a small table, covered with a white cloth bore the family Bible which had seen many years and had been well used. The edges were worn and the cover protected with green baize. Another smaller white cloth was also spread upon the cover to keep off the dust. The clean bed, in which Mark had slept, occupied one side of the room, and a plain bureau with a few hickory chairs completed the furniture of the room.

"You find time for your garden, Druker?" said the Colonel, looking through the window.

"Yes, we both find time for it, my wife and myself, and the children will help pull up and hoe up the weeds when summer comes on. I am going to plant corn also in that field across the road, up to the foot of the hill, and this orchard gives us all the peaches and apples we want the year round. We dry many of them in the sun."

"Plenty of eggs?"

"Yes, sir, and pigs too for ham and eggs."

"Our Kentucky dish," said the Colonel laughing.

"In old Virginia first, Colonel. From there it traveled over the mountains to Kentucky."

"You're right, Druker, stand by your own. But we can't stay, we must travel too, and pretty fast, if we want to be back this way in time to see Mr. Clark."

They both shook hands with Mrs. Druker; then, led by her husband, they soon regained the wagon, and started for Mill Creek. Druker had told them that the way along the creek was too much frequently harbored upon the trees, catamounts like the thick timber road off the Elk which they had visited a few hours before. Anxious to gain time, the Colonel urged his horse to a trot whenever a level stretch along the creek allowed him to do so, but, at each crossing, he had to use caution in leading it in the water to avoid running the wheels against the boulders, while the black slabs of coal which paved the bottom the creek, and the tall trees which shaded their way on either side, satisfied him that he and his friends had indeed made a good investment.

"You will have to survey these hillsides to locate a tramroad, George," said the Colonel.

"Yes, high enough not to be covered by a sudden rise in the creek and washed like these rails which are still lodged in these branches, eight and ten feet above this water. Good many trestles will be needed over the hollows."

"You will have to include in this cost a good sawmill at the works, which our engine and boiler, now there, can run without strain. We had this in mind when we set them up two years ago, as well as for a corn mill. It is a large boiler and a large engine."

They were just passing the two wagons returning from the landing with their loads of empty barrels.

"Mr. Mark all well at the works, my man?" inquired the Colonel, addressing one of the white teamsters.

"Yes, sir. He was all right this morning when we left. Are you Colonel Albert, sir?"

"I am. How do you know?"

"He told us that he was looking for you any day. Mr. Mark is a great worker, sir, and we all like to please him. He does not ask anything but what is reasonable."

"That's right. How long have you been working them horses?"

"Since the works were put up, sir, two years ago."

"This is a hard road for them to pull on to be sure, but can't you make them look a little better? They seem to be of good stock. Have you not corn and oats enough for them?"

"Sometimes we have had, and sometimes not, sir, but, from now, I think, rations will be more regular, all round."

The Colonel and his son exchanged significant smiles. Addressing again the teamster, he inquired: "Do you mean Reynolds?"

"Well, sir, it's no use to name anybody except for good, as Mr. Freenberry often tells us on Sunday, but I think we will have a better chance after this. These poor beasts can't speak their pains, but their bones have tongues. They're always willing though to do what they can."

"What's your name?"

"Andrews, sir, Jack Andrews."

"Jack, how often does this creek rise up above your head?"

"That depends upon the rains, sir. Sometimes two or three times a year."

"Back water, is it not?"

"Yes, sir, when the Elk and the Kanawha are full and rising, and worse yet when the Ohio is full too. I have seen our flat up there on which the works stand covered over one foot, and all the fires had to stop for nearly three days."

"Yes, I remember the time," said the Colonel. "Some eighteen months ago at Cincinnati, the cellars were full of water four blocks from the river, and the gas lights in the lamp posts on the levee were but a few inches above the flood."

"That must have been at sight at night, sir! But our gas lights all went out here, the retorts not going."

"Well, I must push on. Good morning, Andrews."

"Good morning, Colonel."

When, after passing the last crossing, they arrived in full view of the works, the Colonel observed his horse, a knowing animal he thought he was, turning his head steadily toward a newly made trough standing at the foot of the hill on the left of the road, kept full of water running down from a tub above it which seemed to have been filled up there also lately. Into this tub the water flowed from a small iron pipe covered by fresh ground and descending from some spring above on the side of the hill. He had also remarked that the horse had twice stopped after for a drink in the creek about a mile back, but, tasting some of it, had passed on, unwilling to drink any more. At this trough however, he seemed glad to quench his thirst, and was given all the time he needed for it, while the Colonel and his son stopped to study the scene which was spread before the eyes of Mark when, a short time before, he had penetrated to the same spot.

The tub and the trough were not there however at that time, but old Tom had showed Mark how badly they were both needed, and when Reynolds, three days after, had returned from the Court House, his wife had put him to work to redeem his word. Mark had also seen Nep Lewis, the owner of the works above, and, following his example, they had commenced also burning their tar and refuse under stills and retorts.

Looking up along the creek, the Colonel expected to see blazes and smoke issue and curl up out of the heaps of coke wheeled off from the retorts. There were indeed old heaps of white ashes and refuse coal, but no smoke. They had all been carefully levelled down to make room at one end for ashes, and at the other end for well-choked and smokeless coke, both piles being distinct from each other, regularly kept up and easily reached, as he could see, by the men and their wheelbarrows, the latter all freshly repaired and in good working order. They could also see the gas lights blazing high above the retorts, telling thus the pressure of a well-conducted coal distillation unchecked by any blazing leak around the lids which closed the mouths of the retorts. A coal wagon descended rapidly on wheels from one of the two entries of the mine until it came to a standstill above the large pile in front of the retorts. When the wheeler, who was singing a song, called out, "Tom, ticket!", an old gray-headed fireman, leaving his shovel down, climbed up a short ladder to examine how full the wagon was, and then handed to the singer a small sheet iron ticket, which Pole, for it was himself, tied up in the corner of a red rag in the breast pocket of his red shirt. Then, raising his end of the wagon, he dumped down its contents.

As he did so, his dancing eyes caught sight of the Colonel and his son looking at him with a laugh. He stopped his song all at once and pointed them to Tom who ran immediately back of the retorts and gave the news to Mark. He was gauging one of the sunken tanks into which the crude oil flowed from the condensers. As soon as he saw him approaching, the Colonel drove his horse toward the Black hands' quarters, and, getting down with his son, he introduced him to Mark.

"Here I am at your call, Mr. Mark."

"I thank you for your coming, Colonel. When we all thus pull together, everything ought to go right."

"I see you have already made a good start with these heaps?"

"Yes, sir, we now charge but three bushels to the retort every twelve hours. We coke them well and make more oil than four bushels gave before. On Saturday evening, we charge four bushels, coke them thirty-six hours, and do not charge again till Monday morning."

"That much coal saved and no diminution in the quantity of oil turned out every week?"

"You have it in a nut shell, Colonel."

"Well, well, this accounts for so much coal now piled up in front of the retorts!"

"Yes, and, to check this accumulation, I have three diggers out of the bank, and set them to make cord wood from the fallen trees around here, to fire ten of these retorts, instead of using coke. Then we will leach the ashes for potash to take down the tar as in the sample I sent you. Potash is as strong an alkali as soda which I have used on that sample."

"Why this is beating Fieldbell flat as a pancake!"

"How?"

"Well he proposes to utilize for this some of his spent acid at the refinery."

"Does he?" said Mark reflecting, with a smile playing on his countenance, "but he can only purge off a very small portion of the tar with this acid so weakened down."

"Yes, that's true and I told him so, but you can do better and at no cost to us for material, for the transportation to the refinery will be so lessened, just one third, and you give him besides a purer oil to distill, for which he will not need as much acid as he uses now, nor as much acid as he would use even on a partial purging with spent acid. You want an agitator here then for this potash treatment?"

"As soon as possible, Colonel, you have several now dismounted in the yard back of the refinery. The round one would do very well. One of the wagons could go for it tomorrow if it doesn't rain. And I could set it up at once just back of the engine, a good place for it, high enough also to fill from it the iron tank on wheels of which I wrote to you for the landing and the boat."

"All this is first rate. But, tell me, what will you do with all this accumulation of coke?"

"Use it in all the houses, grates, and stoves instead of coal. Every bushel of coal costs from twenty-four to thirty-four cents, and that much also can be saved by using this beautiful coke."

"You have a sharp eye, friend Mark, and you are right every way. Waste not, want not."

"No slothfulness in business, Colonel. You remember the word?" and Mark raised his forefinger pointing it upward [Romans 12:11]. The tone of his voice was lowered and earnest. The expression of his face that of an attentive listener repeating a command.

"Yes," added the Colonel, "and fervent in spirit, serving the Lord."

Andrews, the teamster, had now arrived. Mark told him to unharness the Colonel's horse and take to it to the stable for feed and to brush it, and as they walked toward the retorts, Johnson, who, in the mine, had been told by Pole that the Colonel had come, was seen with his miner's lamp locked on his cap, hastening down the wheelway toward the ladder.

"Here is Johnson coming to see you, Colonel," said Mark.

"Yes, you wrote to me that he does honest work. Here you are, Johnson, do you remember me?"

"Yes, Colonel, I do. I am right glad to see you at last. How is Col. Truesdell? You have not been here for so long, neither of you, but you have come to stay, Mr. Mark says. And if it's all right, Jim, run and tell my wife that the Colonel and another gentleman are here."

"My son George, Johnson."

The rough miner shook hands heartily with the young man and continued: "Tell her that the Colonel and his son will come up with Mr. Mark to take dinner with us." Jim was his own son, a miner, and lost no time in a run across the creek on the wood plank to carry his father's word to his mother.

"Jim," cried the Colonel, "Ham and eggs, nothing else for us!"

"Yes, sir," answered the young man, turning half round and laughing.

"You seem all to be hard at work here, Johnson?"

"Yes, sir, some sort of a new life has come into us all lately," he replied, looking at Mark, "and the more we work, the more we want to work. We can't help ourselves. See all that coal piled up so high. I have never seen the like of it."

"Have you not too many diggers?"

"Well, Mr. Mark has already taken off three to cut wood, and the rest dig no more extras over their eighty bushels, nor have the wheelers any more extras to wheel, but not one of them grumbles. They get better eating, and don't have to do their own cooking neither, since Mr. Mark got Maria to come up for that."

"Who is Maria?"

"The wife of one of the firemen," answered Mark. "She was engaged two years ago with him to do the cooking for the hands, but she did not like it and went back to her mistress. I wrote for her to Mrs. Wright, explaining the circumstances, and she sent her back. I filled up that little cabin for her and her husband, and she is very well pleased now."

"Yes," continued Johnson, "and the men have better clothes too, new shirts and new shoes all round. Their masters are better pleased when they see them on Sundays come in more contented, not growling and fault finding as they used to do."

"It has not taken you long, Mr. Mark, to make all these changes?"

"You had given me full authority, Colonel, and I have used it without loss of time for the good of us all."

"This is the coal measure you wrote to me about, for every charge of coal?"

"Yes, sir, a three-bushels measure for every heap. Here is a four bushels measure hung up on that spike for Saturday evening. These four-bushels charges have till Monday morning to coke, and, instead of two firemen on Sundays, we need but one whose work is very light."

"These lids are tight, none leaks."

"Really?"

"Yes, sir," said Johnson. "Here is Wag, a slave, who could not sleep now if any of them happened to blaze. He is the man who bakes the clay and grinds it fine as flour." Wag was listening and grinning while at work at his mill and showing the whitest set of more teeth than the Colonel had seen for days.

The whistle blew. It was twelve o'clock. All the diggers and the wheelers had already come out of the bank and were seated on the steps and on a bench in front of their quarters, eyeing the movements of the Colonel and the little group with whom he was speaking. The rest of the hands from the works soon joined them and all went in to their dinner, which Maria had spread upon a long table, fastened to the floor of the cabin as were also the two long benches, one on each side. She was a good cook, prided herself on her work, and as Mark frequently called in to see how she was getting along, his pleasant inquiries and advice encouraged her to feel that she must at all times, at all meals, be the same attentive caterer. Her rations for the week were regularly served to her every Monday morning, and Mark personally inspected them to satisfy himself that they were in weight, quantity, and quality what they should be according to the terms of the contract signed by Fieldbell as President of the company with the owner of every hand.

"Let us go to dinner also, Colonel," said Johnson. "My wife has got it all ready." Leading the way on the wale planks across the creek, he went up the hill to his house which stood at a short distance and on a line with one of the two entries into the mine. His wife was on the porch, waiting for them, and, in an off-hand way, greeted the Colonel. She reminded him of the deer meat she had cooked for him and his friends two years before in that same cabin.

"Yes, I remember, Mrs. Johnson," said the Colonel. "We all had a long tramp through these woods, these hollows, and these hills, and were hungry like wolves in winter. That deer meat tasted so good, and you had killed it the day before for us, Johnson."

"Yes, sir, at a lick two miles from here, up the creek. Now, sit down, gentlemen. We have no deer today, but the Colonel said he wanted ham and eggs, and here they are in regular Virginia-style. Colonel, will you ask a blessing?"

With bowed heads and thankful hearts they all united in the same petition. Mrs. Johnson's two daughters stood helping the guests, while, at the head of the table, she sat waiting on them also.

"Mountain air is good for your children, Mrs. Johnson," said the Colonel. "How both your daughters have grown these last two years! I must bring my wife someday to get acquainted with you all. She is a great housekeeper. We have raised two boys and four daughters, all living yet. Two of our daughters are married and all at home in Frankfort now, except George here, our eldest son, whom I have brought up to help me on this journey."

"Helen," said her mother, "bring the pitcher of milk and fill the Colonel's glass."

"You must have as good a cow as Mrs. Druker, Mrs. Johnson."

"This milk, Colonel, is from Mrs. Reynolds's cow which she left here in my care, if Mrs. Mark wants to buy her by and by."

"Oh yes, she told me. How do you like her, Mr. Mark?"

"Very well, Colonel. I think I will take her at her price when my wife gets here, if Mrs. Reynolds will wait that long."

"Yes, I told her that I would arrange this with her for you and pay her the $30 when you say so. Have you seen the horse, George?"

"Yes, sir, one of the men showed it to me on the hill grazing back of their quarters. It seems to be in good order."

"Reynolds told me, Mr. Mark, that he wants one hundred and ten dollars for it and the saddle, but that you have no need of it at present. I said to him that George would need a horse for the present, and would try it, and that, if he liked it, I would pay him the one hundred and ten dollars. If you want it any time hereafter, you can have it of course at the same price. George will take good care of it."

"Very well, Colonel. Everything in its time and order."

"You are all Union men, I see," said the Colonel, spying a late number of *The Cincinnati Commercial* on the Bible upon a side table.

"That's what we are, Colonel, in these mountains, and nothing else," answered Johnson, "except a few block heads who can't be reasoned with, no more than most of those lawyers and land owners from old Virginia at the Court House."

"Yes, I have seen them," said the Colonel, "a hard set to talk with but I think they will stand by our Baltimore Convention on the 19th of this month of May, our Constitutional Union Convention."

"That's ours too," said Johnson. "But Mr. Mark, I think, favors the Chicago Convention."

"Yes, the Republican National Convention which meets on the 16th, three days sooner. That is, if the nominations are what they should be," replied Mark.

"I thought you was a Republican, Mr. Mark," said the Colonel smiling. "From your whole make up, you are very radical in your mind and your ways, fond of digging to the root of everything and finding out what it is."

"Yes, Colonel, no search, no find, but no compromise ever with malice, error, or conceit, when the well-being and the union of our country hang in the balance. God gave it to us to keep one and undivided for our children and transmit it to them as we received it from our fathers. Slavery is one of the large parasites that have gradually grown upon our nation as excrescences do grow day and night. They suck our life's blood and should be cut off."

"You are right, sir. No yielding to malice, error, or self-deception, as you say, nor to any such parasites, for it would only end in the ruin of our country. Well, we shall hear in a few days who the nominees will be to help us out of the mess. Have you seen the Lewises above since you have been here?"

"Yes, sir. I found that they were throwing into the creek all the refuse of their refining, and went to remonstrate with them that our teamsters were compelled three times a day to take their horses up the hollows to distant springs to water them, and that they could easily leave the creek free by burning all their refuse with their coke under their retorts and stills as we had begun to do ourselves here with what tar we now get from the settling tanks. They demurred a little, but I insisted that this running water from spring to mouth was public property to be kept clear from damage and injury, and invited them to come down and see how we burn our tar now in our clay troughs under the retorts instead of using coke, just as we intend to do as soon as we take down here in an agitator all the tar from the crude oil which we make every day, namely one third. They opened their eyes very wide but made no more objection. The Doctor, who is the eldest, came down the next day to see how we do, and went back satisfied that the change would pay them. Since then, the water has much improved. In a little while, after a good rain, it will taste still better, and, after a few weeks, it will be as good as it ever was. They were very curious about the way we would adopt for this purging, but they will have to find out."

"We have a controversy with them about our dividing line on this very hill," said the Colonel.

"Yes, Reynolds told me something about it. Mr. Fieldbell contended, upon what Mr. Johnson has reported to him, that they have passed the line and are now digging coal that belongs to us on this hill."

"Yes, we must have another survey made of the county surveyor as soon as possible. We are neighbors, and cannot afford but to be at peace together. I will see the surveyor tomorrow and make an appointment with him. Our deed is in the safe in the office. He will bring it along and compare it with theirs. Please tell the Doctor, Mr. Mark."

"I will, sir."

"You know, Johnson, where was this vein of cannel coal first discovered here on this hill between the two entries?"

"Among the roots of that old black oak just above the creek."

"Yes, the same tree that old White was going to cut down for the bark, when his eye caught sight of the coal among the roots. Well, that's our place of beginning, and our deed specifies the exact distance from there to the dividing line. Perhaps Lewis's deed does not state this as ours does, and they or we may have to fall back on the parties from whom we bought. By the by, Johnson, we will have also to survey your digging as far as it has gone to the present time and make a map of it showing the extent of the mine on the other side of the hill, with all your rooms and pillars which will have to be numbered."

"We have been speaking about it, Colonel," said Mark. "When the surveyor is here with his instruments, we will get him to give us all the points, that we may know their position from the hill outside over the root of the bank, to locate one or two air shafts now much needed to ventilate the rooms."

"I will tell him, Mr. Mark," answered the Colonel, "that he may bring what instruments he must have for this underground survey. Then he can give us a current map of it, and plant the stakes outside on the hill corresponding by numbers with the stakes inside."

"The size of each room and of the pillars between," said Mark, "will show us, I fear, Mr. Johnson, that, in many of them, the diggers have left them and gone off to others, before taking out more coal than they have actually done. The sooner this is stopped the better for us all."

"That's true, sir. I am anxious myself that all them measurements should be made, that we may know what we are doing, and whether also Lewis's have passed our line."

"Mr. Mark," inquired the Colonel. "What do you think of that petroleum in Pennsylvania as a competitor against our oil in the market? There is no tar in it."

"It is indeed a very clear natural oil, Colonel, and easily, cheaply refined, but half of it nearly is too light and dangerous to burn in a lamp, or even to hold against any heat, for it explodes so readily that it cannot be used, this half I mean. It also forms a waste, at least so far as we now know, while the heavier portion burns off so rapidly from the wick that our heavier and steady burning coal oil will, I think, remain the favorite, as it is now."

"What competition there is, however, will compel us to lower our prices?"

"Yes, undoubtedly. Hence, let us save wherever we can."

"Right again, Mr. Mark, you are indeed a first-rate radical, and you may yet bring me over to think as you do of your Chicago Convention when the nominees are out. By their fruits we shall know them [Matthew 7:15]. Now, Mrs. Johnson, you have treated us very handsomely. That deer of yours, Johnson, two years ago, was very fine, but your wife's ham and eggs and nice biscuits beat today the deer all out, for I felt very, very hungry when we came, even after feasting on Mrs. Druker's cold biscuits and her rich milk. I told her that we would stop, on our way back, to see Mr. Clark and his school if it was not out yet. We will have to start shortly to be there in time."

Rising from the table, followed by his son, and by Mark and Johnson, he shook hands with Mrs. Johnson and her two daughters, and, on the porch, at the door, stood a few minutes looking down at the works across the creek, and beyond them at the cabins on the hillside opposite. Pointing with his finger to that formerly occupied by Reynolds, he turned to Mark, while going down, and said: "That is the cabin in which you now keep bachelor's hall, Mr. Mark, till your wife and your little ones come up to join you in it and help you on your good work?"

"Yes, Colonel. Mrs. Johnson has very kindly taken me in to board me till then."

"At what time do you eat breakfast?"

"At half past seven o'clock, after the retorts have been charged and the coke has been wheeled off."

"And supper?"

"At seven again after the evening charge."

"And you go to bed at what time?"

"Between nine and ten."

"Your hours and ours are about alike and we can thus everyday think of each other ten miles apart."

"A good thought, Colonel, to bind together in heart friends and families separated by distance."

"You hear often from Mrs. Mark?"

"She writes as soon as she gets a letter from me, and I do the same."

The Colonel's voice was slightly choked, when, as if to hide his emotion, he quietly added: "Oh! You must soon be together again!"

"We will, Colonel, as soon as she can safely undertake the journey."

The Colonel stopped after they had again crossed the creek, and, examining a space under the building, between the retorts and the engine, he said to Mark: "This is where you propose to set up the agitator?"

"Yes, sir, under this large beam."

"And the saw mill of which we spoke in Cincinnati to cut the timber for the trestle work of our tramway when we get at it?"

"Well, on the opposite side of the building and on this same line with the agitator. The same pulley for the two belts."

"Yes, this is the right place for both."

"Father," said George, "I will go and saddle Selim."

"Yes, you take him along, and begin our trial. We will try him also in the shafts some other day."

"He is a safe gentle horse, Colonel," said Johnson, "to ride or to drive, either way."

"Frank," said Mark to the Black man whom Reynolds had trained as his assistant engineer, "are both wagons loaded for the landing?"

"Yes, sir, there they are, ready to start."

"Mr. Mark," inquired the Colonel, "you would simply band and bolt to the beds of these two wagons the two iron tanks to carry the clear oil to the boats at the landing after filling them here?"

"After throwing off the tar, Colonel, we will not want these two wagons, nor two iron tanks. One only will do to make two trips a day as now, and it will thus carry off the daily yield of these one hundred retorts."

"So that we will need only one teamster and one team?"

"Yes, sir."

"We can then dispose of these two horses, save their feed, sell the barrels also, and use the money from the tank?"

"Yes, and for another boat tight and decked over to be filled, after a few trips, by a pipe down the bank of the landing from the contents of the tank. The boat now in use will have to be made tight also and decked, to take the place of the other when filled and ready to go down."

"What of the chill on that oil in winter in tank and the boat?"

"We will tank it here hot from the agitator, and, in winter, will pack the tank in saw dust, so that, at the landing, it will yet be warm enough to run down the pipe into the boat. At the refinery, a steam oil can easily be used to warm it again and fit it for pumping from the boat up into the stills."

"This trip has paid me well, Mr. Mark."

"Oh! We will succeed here, Colonel, as long as we all thus pull together."

After another walk along the retorts, and examining how the tar was emptied by one of the firemen from a pail into the new funnels whence regulated its flow down by a stop-cock into each clay trough under the refinery, the Colonel, still with Mark and with Johnson, reached his wagon by the side of which stood George on Selim. He had slung the strap of his rifle over his shoulder and had already tried the gait of the horse.

"How do you like it, George?" his father asked.

"So far very well. He is gentle and paces too. Mother, I think, could ride him without any trouble."

Observing a smile on Mark's face, the Colonel told him, "Mr. Mark, my wife is but a little younger than I am, and yet she can still ride very well. She was once one of our Blue Grass girls."

"Oh, well, I understand, Colonel."

"That's where I was born and raised too, Colonel," said Maria the cook, who stood at her door washing her plates. A broad smile lighted her honest face.

"Is Maria your name, old woman?" inquired the Colonel looking at her.

"Yes, sir."

"Mr. Mark spoke to me of you and of your good cooking. How did you leave the Blue Grass for this Kanawha country?"

"After Massa Todd's death at Versailles, Col'nell, my young Miss Maria, his daughter, came to the Licks with her husband Major Wright, and brought me along with them. That's many years ago."

"Why, my wife was a Miss Todd too. Her father was Mr. George Todd, of Lexington."

"My old massa's name was Harry Todd, the cousin of Mr. George Todd. I remember well Master George Todd with his red hair and red mustache."

"Well, when you see my wife, you may perhaps both recognize each other. You seem both to be of the same age."

"I think I know her, Colonel. Was not her name Harriet?"

"Yes, it is yet, Maria. Why she will be glad to see you once more."

"And won't I, Master Albert! I am so glad you come up. Make me young again, as I think of my mother and how I used to skip about!" Tears came up to her eye as the Colonel took her right hand for a friendly shake.

"Where is your husband, Maria?"

"There he is, Master, looking at us," she replied and pointing her finger to old Tom the fireman who stood off at the corner of the building with his shovel in his right hand, and as if he were riveted to the ground by the sight of the Colonel shaking hands with his old Maria. His rags had given place to a new suit of working cloths. His long feet were also well encased in a new pair of substantial shoes, and an ample felt hat covered his gray hair.

"That's all right, Tom," cried the Colonel to him. "We won't carry her away. We are talking of old times!"

"And where are your children, Maria?"

"With Missis Wright, Col'nell. Three daughters. One married and two grown up, all living together with the husband of the married one, a fireman for Major Wright who has a big salt furnace there."

"Well, Maria, Mr. Mark has been very glad to have you cook for the boys and so I am especially since I know where you came from. Good-bye, we must be going. Tomorrow, then, Mr. Mark, you will send one of the wagons for the agitator, the cog wheels, and all the iron work that belongs to it?"

"Yes, sir. It will be there tomorrow at about ten o'clock, and we will go to work at once to set it up. I will also have the cord wood hauled up to burn under some of the retorts and will stave up half a dozen leach cribs to throw in the ashes and get the strong potash water out of them. I have a big kettle here, a big huge kettle, which I will brick up to boil down the lye."

"You have to boil it down?"

"Yes, I worked a long time in Chicago before I found how strong the lye should be, either of soda or potash, to purge off the tar effectually. Though I had very little light oil there, for lamp use, remaining after the tar had been thrown. It was the tar and heavy oil, and ammonia water from the gas works, not the product of cannel coal distillation like this oil. The coal was bituminous, from Pennsylvania, such as is generally used for making gas."

"I will tell Fieldbell to have the agitator well cleared up and in readiness for the wagon as soon as it comes. Now, we must go. If you want anything else to come up in the wagon, anything light, I mean, for that agitator weighs over a ton by itself, does it not?

"Yes, about twenty-five hundred pounds. The bulk, six feet in diameter and six feet deep, will be more difficult to manage than the weight, and lest any accident should happen on the road, it is so narrow and so rocky in some places, I will send two of the three diggers down with Andrews to help him."

"I could come up with them too, father," said George, "and would still have time to ride back before dark."

"Yes, provided no catamount springs down on your back," said the Colonel seriously.

"You take the open road on the right, Colonel," said Johnson, "nearer the Elk. Leave the timber on our left. Those varmints don't come into the open."

"We will do so, Johnson. Well, good-bye till I come again not long hence. In the meantime, George will ride up two or three times a week and will carry the mail. He will bring you the news from Chicago and Baltimore. Good-bye, Maria." Bowing to them all, he shook the bit and started his horses following George who seemed well pleased with Selim's gentle and quick ambling gait. Both horses seemed also well pleased to find themselves in each other's company.

"Father," said George. "This Mr. Mark has all the look and ways of a first-rate man."

"He has indeed. The more I see of him, the more I like him. He is a Christian gentleman, one of the kind that wears well."

They soon overlook the two wagons, and, after exchanging a few words with the two teamsters, Andrews and Lewis, they pressed on to reach the school while it would still be in session. They arrived in time. Mr. Clark was offering a closing prayer for a divine blessing upon the studies of the day, and, when he had ended, seeing the two gentlemen open the door and enter, he went to offer a chair to each near the door, and beg them to wait until the closing hymn of thanks had been sung. The school was large, and, under his gentle lead, the Spirit song was so well rendered that both the father and the son looked at each other in amazement and delight.

"This beats all the school singing I have heard in Frankfort and Cincinnati," said George. A short silence followed the singing, and as Mr. Clark and the scholars still stood up, the boys on one side and the girls on the other, all looking steadily at him, he pointed his finger to some of the latter who occupied the bench nearest to the door, and, at the sound of his voice calling "Mary!" she led off to the door, all the rest of the girls following her, bench after bench, each with a book or two in her hand, after which "Henry!" was also called. He stood nearest to the door, and he led off as Mary had done, quietly followed by the boys, bench after bench and carrying their books, until only Mr. Clark and the Colonel with his son were left.

"Col. Albert," said Clark, "Mrs. Druker sent me word at dinner time by one of her little boys, that you and your son intended paying us a visit on your return from Mill Creek this afternoon. I am glad, gentlemen, to make your acquaintance."

The Colonel had been informed of the character of this school by Mark in one of his letters but was turning in his mind whether even the half had been told him. Taking the hand which Clark offered him, he replied: "You are engaged in a good work, Mr. Clark, and I am glad also to make your personal acquaintance. Although I had already heard of you, all the way down at Cincinnati through our good friend, Mr. Mark."

"So have my friends in Ohio heard of him through me, Colonel. We have both enlisted in the same war against the world, the flesh and the devil. I know at least I feel almost certain that you have come up here to fight with us for the redemption of these poor people. They realize how little they and their children know and are determined to work to make this schoolhouse all it is worth. I am just as much determined also to help them as long as they want me."

"You teach ten months of the year, I understand?"

"Yes, sir, six hours every day. The progress which these boys and these girls are making is truly astonishing. Look at those two blackboards." On one were geographical outlines, and the names and capitals of some of the states of the Union, also the degree of longitude and latitude numbered and drawn across and vertically, evidently the work of two or three different children. The other was crowded with arithmetical figures of additions, subtractions, multiplications, divisions, and fractions, all carefully chalked, stood in their regular positions. They were the only indications left that a school had just been held in the place, for not so much as an inkstand, not a pen, no paper, not one book were to be seen anywhere on the benches, and these were all ready for a religious service without having to be moved in line.

"No Bible here, Mr. Clark?"

"Yes, sir, here it is," answered Clark, drawing the small book out of his pocket. "I teach them to memorize it and some can repeat already by heart whole chapters of the Old or New Testaments, studying at home in the evening."

"Thus was I taught it, Mr. Clark, and it has remained ever since with me an endeared memory, an ever present consolation, the safest guide I could desire through all duties, through all trials."

"This is also my experience, Colonel, and has been from the time I was a boy just as these are. Having freely received, we cannot but freely give also" [Matthew 10:8].

"How far do these children live?"

"Most of them within two miles. A few of the oldest however don't mind walking three and even four miles through all weathers and all kinds of roads. There is a little run right back here in which they sometimes wash off the mud from their feet or their shoes before they come in, boys and girls. I love to teach them, and they love to learn."

"How often have you preaching services in this room?"

"Mr. Freenberry will preach here next Sunday. He comes to us three times a month and preaches in the morning. It was upon his writing to me about these people eighteen months ago that I came here. He looks upon this field as his best and most promising one."

"I will certainly try and be here at his next appointment three weeks hence," said the Colonel, "for I want to make his acquaintance also. I will come with my wife."

"We will be heartily glad to see her with you, Colonel. It is thus by working in unison, hand in hand together, that Christians can hope to obtain the blessing of God upon their attempts to have their own knowledge of Him with those who know Him not."

"Father," said George pulling his watch. "It's getting time for us to go."

"That's true. We are strangers here, Mr. Clark, upon a strange road. Please tell Druker and his wife, if we do not see them in passing their house, that we had to hurry off in that account. They will understand us."

"I will, Colonel, but I think that they and their two little boys, who were here a few moments ago, will be watching for you from their porch. Good-bye." He stood at the door of the schoolhouse, watching them as they went. George at a rapid pace preceded his father as before, and, as they reached the orchard in front of Druker's house, Clark could see him take down one of the bars of his gate, and, with his two boys and two dogs at his side, shake hands with the Colonel and his son while they waved their hands also to his wife. She returned their salutations from the front of the house.

The sun was going down when they passed the refinery. Without stopping, they took their horse to the livery stable which was connected with the hotel, and arrived at the house a few moments before the usual hour for supper.

Mrs. Lorme's mother, like Mrs. Albert, was a ripe Christian mother, but her daughter and Mr. Lorme were that clan of still, intellectual professing Christians, self-dependent, who think but little of and realize still less their daily obligations to Divine Providence for pleasant surroundings. And Fieldbell, who had, like themselves, been trained up also by Christian parents, little thought as they did of the value of his birthright. These three however deferentially bowed their heads, when upon taking their seats, Mrs. Lorme, as she did before every meal, requested the Colonel to ask a blessing upon it. The conversation soon turned upon their day's journey.

"That scenery in the Elk valley," said he, "till the road along the foot of the hills turns off at the mouth of Two Moles' Creek is really beautiful, but my admiration for it soon subsided after we had entered the timber road from there to shorten the distance, as we thought, when we should rather have gone on the old flats on the left. Still, I felt much impressed in that timber cutting with the great wisdom of this counsel of Solomon 'a soft answer turneth away wrath, but grievous words stir up anger' " [Proverbs 15:1]. Then he detailed their encounter with the catamount, its fierce look and growl, and their soft parting salute to him in answer.

"That was indeed well done," exclaimed both the old ladies almost simultaneously, while Lorme, his wife, and Fieldbell smiled at George, whose serious countenance however seemed emphatically to endorse his father's sentiment, for he looked at them in response and said: "Yes, I first discovered it, and my rifle of course was my first thought, but I had a glimpse of its length through the branches, and, if it was not eight feet from head to tail tip, it was not much less. Indeed, I felt thankful it did not jump after us. Our horse too was glad to skip out as cheap as it did."

"We call them 'Loo-sevee' in Vermont from the Canadians who name them 'Loups-cerviers,' " said Mrs. Lorme's mother. "Our farmers there, I remember when I was young, had all they could do to keep those fierce beasts from carrying off their sheep and killing their cattle."

"Yes," added the Colonel, "I am satisfied that I acted all right. I have a whole skin yet, and you young folks are so valorous just here!"

"Oh! Colonel," remonstrated Mrs. Lorme, "I would have done worse than you did. I would have shrieked and jumped out, and run, and gone on so that I am sure my craze would have brought the panther upon me, and I would had been torn to pieces."

"Glad I am then that you was not with us. George kept cool as a cucumber. He realized that, rifle and all, we were hardly a match for the varmint, as Johnson at the mine called it."

"How do you like Johnson, Colonel?" inquired Fieldbell.

"I think him a good faithful miner, sir. I am glad he is there for us, and that he listens so readily to Mr. Mark's suggestions. I must see the county surveyor tomorrow and have him go up, not only to survey our line again to settle it with Lewis, but to survey our mine as well, its rooms, pillars, and entries, to map every one, and ascertain how much coal by the cubic foot we have taken out so far and how much remains in the pillars for instance, then how much we can expect out of that bank till we reach the other side of the hill. Johnson will then be able to locate the two or three air shafts which he says he needs badly now to ventilate his rooms. The smell of the powder stifles them sometimes, and they have to use so much of it, for the coal is getting harder and harder the deeper they get into it."

Fieldbell was listening with close attention. He never had mentioned these facts in his correspondence with the Colonel.

"Did you see Lewis, Colonel, about our line?"

"Oh, no. The surveyor can settle that with them. I will let him take our deed, so he may compare it with theirs, and, when he comes back, he will tell us all about the difficulty that it may be arranged at once. George will go with him. We are close neighbors there, and as Mr. Mark says, we must be neighborly too, live at peace with them. He has already brought them over to his way of thinking and doing in regard to keeping the creek clear of the tar and refuse from their refining."

"How?"

"Why, by burning all this tar and refuse under their stills and retorts, as he does under two of our retorts with what little tar settles in the tanks from our crude oil."

"How does he burn it?"

"In clay troughs which he shapes out of fire clay, then roasts in the oven, and then lays flat upon the grate bars. Instead of using coke, he feeds his fire in these troughs with the tar which he runs down from an iron funnel outside bearing upon a stop-cock on a sort of iron goose neck which reaches into the troughs through the front brick wall above the doors. He says that you could do the same here, Mr. Fieldbell, under your stills. Don't people who have wells near the refinery complain of our well there into which our refuse runs down?"

"Yes, to some extent," replied Fieldbell hesitating.

"We must fill that well, Mr. Fieldbell, but first burn the tar and refuse now in it, some night when the wind can blow the black smoke from the town toward the hills. Then we will get clay troughs too and try to win back the good will of all our neighbors. They could sue us for damage. I would if I were in their shoes." His honest soul revolted at the idea of imposing upon the good nature of his neighbors.

"I often thought of the same thing, Colonel, that it could be done."

"Well, you did not know how, I suppose, but it works very well, I assure you, for I have seen it, and then when we make that change, when we throw down all our tar at Mill Creek, see what quantity of beautiful coke we can lay by to sell here and down the river when we have our tramroad in operation. But it's a long story. We have much to talk about tomorrow. Mark's coke, by the way, as he wheels it out on a big pile by itself, is so well choked that none of it takes fire in that pile. The ashes he also wheels out and heaps up by themselves. There is much method in his ways it does my heart good to observe how intelligently he goes to work and how well he loves to work. But I am getting quite enthusiastic about Mill Creek and Mr. Mark and well may I, when I think of Mr. Clark's school at our landing."

He went on describing in detail his visit there and conversation with Clark. "Why, Druker too, the boatman and his wife, you should see what a worthy couple they make and how true to their two little boys! Had we not a pleasant trip today, George?"

"Pleasant every way, father, save that catamount."

"You will dream about it, George," said Mr. Lorme.

"I would not be surprised, sir, if I did. It was ugly enough to stamp itself on my brain and bring on a dream."

"Mr. Fieldbell," said the Colonel. "There is a Black woman cook at Mill Creek, belonging to Mrs. Wright, whom you hired with her husband, old Tom, to cook for the hands?"

"Yes, sir, and she had not been there, but a few months when she took it into her head to go back to her mistress and go she did. I went to see Mrs. Wright at the Licks at the time, and she told me that Maria complained of the rations that Reynolds was giving her that she did not get as much nor as good as they ought to have been. Mrs. Wright agreed to charge only for the time they had served, and I so settled with her."

"And the men went on cooking for themselves?"

"This is what Reynolds told me and that they were satisfied. But as soon as Mr. Mark accepted your proposal, he wrote at once by old Tom to Mrs. Wright, who brought to me his note requesting her to send the woman back on the same terms as before, and to come to me to sign the arrangement till Christmas. Tom was with Mrs. Wright and said that they all wanted Maria again since Reynolds was going away. I signed it of course."

"Well, she is doing first rate now, and appears well contented, so do the men, especially old Tom. Mr. Mark drops in now and then to taste of her cooking and speaks highly of it."

"He does!" all exclaimed with surprise.

"Yes, he does, for his eyes are everywhere. He declares she cooks splendidly and is as tidy and clean a creature as one can be about her dishes, plates, and knives and forks, even the tin cups are scrubbed and shine as bright as silver. She comes from our Blue Grass, Hattie," he added addressing his wife, "from old Versailles, where she was born and raised on Harry Todd's farm, whose daughter Maria married Mr. Wright who owned a salt furnace here. After their marriage, they brought her along to live with them."

"Why, yes, I remember her very well," replied Mrs. Albert. "She was a smart girl."

"Tears gushed up to her eyes, when I told her that no doubt you would remember her, and that you was here and would come to see her. She could hardly let go of my hands."

"Yes, indeed, I would like to see her again. Many a run we had together, many a climb too in the orchard after cherries and apples."

"Mother," said George. "I have a very gentle horse for you to ride on to Mill Creek, a real lady's horse, pure Virginia blood." The Colonel here stated his unconditional purchase from Reynolds and arrangement for George with Mr. Mark.

"I have my saddle here yet, Mrs. Albert," said Mrs. Lorme. "If you are not afraid and can stand these two long rides in one day."

"I think I can, by making an early start some fine morning."

"Yes," added the Colonel. "The Reverend Mr. Freenberry, the preacher, whom they all think so much of there, preaches in the schoolhouse three times a month. He will preach in it next Sunday, and again the third Sunday following. Maria will be sure to be there, and while I think of it, we can spend the night at Druker's, and the next day go up also to see Mr. Mark at the works. So that, between Mrs. Druker, Maria, Mrs. Johnson, and all their children, you will have a good time too."

"What about the catamount?"

"Oh, we will give it a wide berth. We will go up on the flats, no timber there."

Thus the conversation rolled on during and after supper. When the Colonel saw Fieldbell go out to retire to his room, he followed him, and, closing the door, he said: "I want to speak to you about some work tomorrow, Mr. Fieldbell. Mr. Mark will send Andrews and his wagon early tomorrow morning, with two of the hands, to help him haul to Mill Creek the round agitator now in the yard back of the refinery, with the shaft, the blades, the pulleys, and all that belongs to it, that he may set it up at once in the place which he showed to me near the engine, and start on his process."

"Well, this is quick work," said Fieldbell, surprised. "How about his soda?"

"He won't buy any soda. He is going to use potash. He has already I don't know how many cords of wood cut up to fire some of the retorts without using coke and make ashes to leach and yield the lye. Potash strong enough to use instead of soda. You may well open your eyes. He is a pusher."

"How did he get that wood cut up?"

"He only charges three bushels of coal to the retort, instead of four bushels as Reynolds did, don't charge on Sunday, and makes more oil now by the gauge than the four bushels Reynolds gave Sunday and all days. Since he has made that change, the coal pile in front of the retorts has got so high that he has taken three diggers out of the bank and set them to cut into fire wood, getting a number of fallen trees out of the way along the creek."

"That's true. Potash is an alkali as much as soda is, but how strong does he use it?"

"Oh, that's his lookout, but he wants the agitator as soon as possible to send you clear oil to work on in your stills."

"Don't he want fixtures to leach his ashes and boil down his lye?"

"No, he is making leaching cones out of staves and has a big hog kettle there which he will set up to boil his lye. That's all he wants his three diggers to give him on that, all the work he needs. Indeed he will want them only for a while to lay up some stock of potash lye. There will also be a teamster not wasted, Lewis. Andrews he will retain."

"How so?"

"Well, by keeping all his tar there and burning it up, he will have one third less weight to send to the landing, and the two thirds remaining he will send in an iron tank on wheels to be emptied by a pipe into the boat. We will need two boats. When one is full, Druker will bring it down for you to pump it up empty and fill your stills. All the empty barrels, with the team and wagon, we can sell to pay for that tank and the new boat."

"Won't that oil in winter chill in the tank and the boat?"

"You can steam it here in the boat with a dry coil, as he calls it, with the bottom of it connected to your booster, so that you may pump it, and he will see to it that it reaches the landing warm enough yet from the agitator to run down into the boat. That will save all your work of cabling up the barrels in those cars on the incline and steaming them one by one empty over the tanks in the ground under the refinery. But I will not keep you up any longer. We will talk more about it tomorrow. Andrews is to be at the refinery at about ten o'clock. Good night."

"Good night, Colonel." Fieldbell closed his door and sat down puzzled and amazed, pondering upon the astounding information just imparted to him by the Colonel and in such a way, such a manner that even if he was President, it was his plain duty to do as he had just been told to do because it was all right and correct. For were he to interpose any objection, he would find himself playing a silly part and a costly one to himself. The motives which actuated him and the ends he had in view were of the dwarf order, selfishness permeated them all to poison the springs of his life in his relations as an employer and as an employee, and even as a husband and a father, for his long separation from his wife and daughter did not seem to weigh heavily upon him. They lived in Ohio. She had a small income of her own and every two or three months he remitted enough to keep her and her child in moderately comfortable circumstances. They were a cold-hearted couple living for this world, rarely thinking of the next and of its realities.

He has not even said a word of my waste acid process, he thought also. But the next thought followed: true, I took the hint from that alkali process. Well, they have me in a corner, and yet they need me here as he is needed ten miles away. I must behave myself, acquiesce, and make no trouble.

The next day, shortly before ten o'clock, Andrews appeared with his team and wagon. The two miners accompanied him. At noon the agitator and all the casting which belonged to it had been loaded and fastened with long ropes upon the bed of the wagon, and George, on his horse followed to go only part of the way and see that Andrews drove cautiously, his two helpers holding firmly on one side the rope ends which dangled from the top of the agitator wherever the narrow road leaned too abruptly toward the other, so that, in such uneven places, the large diameter of the iron tank threatened at times an over throw. Andrews drove slowly. When they had arrived within one mile of the landing, it was nearly four o'clock, and George, satisfied as well as the men that they could get along without him, turned back. In one hour, he was at the refinery with his father again.

To the latter, Mark had sent by Andrews a written request for his mail enclosing another letter for his wife, and the Colonel had fastened to the front bed of the wagon a small leather pouch which he had made that morning, with a brass lock on it, and in it he had enclosed the two latest numbers of *The Cincinnati Commercial*, also a weekly religious paper from Philadelphia. The key of the lock he had also fastened, for safe transit, by a short piece of twine, round the neck of Andrews, who, at first, on the way, frequently felt it with his hand to assure himself that he had not lost it, until the thought occurred to him to request George, as they were taking a short rest, to untie the string and fasten it again round his neck under his shirt collar, so that he could feel the key hanging loose upon his breast.

In his letter to the Colonel, Mark had stated to him, and the Colonel had read it to Fieldbell, that, as he had now intimated it to him, the firemen were doing so well, it was his intention shortly to increase by one third the present production of oil, by charging the retorts every eight hours instead of twelve hours. In which case, if the yield turned out as he hoped it would, the chargers would be monthly paid an extra wage, and Lewis and his team would probably be needed only part of the time.

"Yes, indeed," the Colonel had said, and answered in a note by Andrews. "We will pay the extra and keep the team. For, as Mark had written, his fires are so well kept up and his three-bushels charges are now every time so thoroughly coked dry that I have almost no doubt, he can in eight hours obtain the same coking, the same yield of oil, and thus realize our wish to increase our crude oil production without additional retorts."

This was another idea for Fieldbell, and he could not but see that Mark was the very man to make it succeed. His own personal labor at the refinery, which the late increase of production at Mill Creek had already also added to, would naturally grow still more upon him as the Colonel had told him among his first words on his arrival from Cincinnati. But it was inevitable. He must refine all the oil sent from the retorts, and his only consolation was the reflection that this oil, gauged of all its tar, would most probably be easily distilled, his labor there after much would be less than it was now.

"Mark my word," the Colonel had told him. "You will not need so many hands here to run your stills. Even if you turn out more of burning oil for Cincinnati, you will only have more barrels to fill and to ship, and I will be glad when you will need more helpers for that. That's what we want."

After dinner the Colonel, with the deed for the land in his pocket, went to the county surveyor's office and explained to him at length the work which he wanted from him, to settle the difficulty with the Lewis's property above theirs and to map out the present condition of the mine and direction of each entry with the size and number of each room and each pillar. He told him that his son, George would go with him the next day, and that Mr. Mark would give him two men to carry the chain and stake it, if, in order to ascertain the location of the dividing line between the two properties, it became necessary to survey again, and find in the woods the old stakes marking the meter and bounds specified in the deeds.

After leaving the surveyor, as he was walking down the main street to his refinery, he observed that a brisk wind was blowing from the river.

This is the very breeze, he thought, we need to carry to the hills away from this town the black smoke out of that well. He hastened his step and soon reached the building.

"Fieldbell," said he, "let us burn that tar and waste oil in the well. The wind blows now from the Kanawha toward the hills, and the smoke will hurt nobody."

"It will last some hours, Colonel, for there is a big pile of tar down there."

"Well, I hope the wind will last long enough to rid us all of the nuisance. Get two or three men to bring their shovels full of hot coals and throw them down." The word was given. The hot coals went down, and the men had hardly, on a run, reached the fence which surrounded the large field when a thick black cloud of smoke rising above a fierce blaze issued from the deep well, and soon assumed such large proportions, while flying toward the wooded hills half a mile distant, that a crowd of men, women, and children rushed out from the streets and alleys which ended upon that rear edge of the town, and stood wondering at what had happened. Every door, window, and low roof had its occupants straining their eyes in the direction of the fire, and it was not long before the word passed from mouth to mouth that the old Colonel had determined upon making an end at once of what he said should never have had a beginning.

"Tomorrow," he added, "Mr. Fieldbell says that when the well has cooled down, he will also throw into it five or six barrels of slaked lime to kill what acid may still remain in the bottom, after which we will shovel back and level down all that heap of ground which has been dug out of it." The owners of the wells around, mostly of the poorer class, who had repeatedly complained to Fieldbell of the injurious taste and odor communicated to their water by the tar, oil, and acid thrown into the well came to thank the Colonel. He apologized feelingly to them for the mistake which had been made and thanked them in turn for bearing so long and so patiently with it.

"Here after," he said, "we will burn this waste every day under the stills, and it will not trouble you anymore." While the masses of black smoke were curling up thicker and thicker, he went into the refinery and heard Fieldbell giving directions to one of the still men to make a thick clay mortar and bed it over the grate bars of one of the stills, then lay fire bricks upon it on the sides and ends until regular clay troughs could be had from Mill Creek to burn the coke from the stills and the waste from the agitators. The blacksmith also had been set to work preparing the iron funnel and syphon to connect them through the brickwork of the front wall with the shallow brick pan thus laid over the grate bars.

"That's right, Mr. Fieldbell, make all things ready to burn up this refuse. We will soon be able to handle it as well as coal. You will give me the size of your grate, and I will write by George tomorrow to Mr. Mark to have his men shape and bake half a dozen clay troughs."

"After he has made his arrangements to purge down all the tar up there, our refuse here will be very much reduced in quantity," answered Fieldbell.

"I know. He will be heartily glad of it, but still he is going to increase his production by one third. Who knows but he may get to double it!"

"Perhaps he may, Colonel."

"And that without a dollar more for more retorts! That would indeed be a feather in his cap!"

Fieldbell kept silent, but the Colonel could see that he did not feel comfortable and turned back to go out and see what progress the fire had made. It was getting late in the afternoon. The smoke and blaze seemed to have increased in bulk as the heat had penetrated down into the mass, and the wind which had also grown higher and stronger gave timely assistance in driving off from the town the rolls of black smoke from the pit which, one after another, arose mixed with forked tongues of fire. A number of citizens had come down to the scene from the upper part of the town and stood around the Colonel watching with him for any deflection in the direction of the smoke, but there was none, and they congratulated him upon his success. The mayor of the town was among them, and he informed the Colonel that this nuisance to the neighboring wells had lately threatened further extension and been quietly discussed in court for abatement.

"I will remain here, gentlemen, till it's all burnt up, all night if it lasts that long. I do hope this wind will stand by us as long as we need it."

Lorme was there also. When supper time approached, he went home with the word that as the tar would probably burn till late in the night, the Colonel had made up his mind to see the last of it and keep all hands with him in case the wind should change to carry the smoke where it should not go, and that some that supper be sent to him as well as to Fieldbell.

At ten o'clock, after burning seven hours, the fire stopped at last and the smoke ceased. The force of the wind had also gradually decreased, but it was still blowing from the river. With thankful heart, the Colonel dismissed the men, and, followed by Fieldbell who carried a lantern in his hand, they made their way to the house. The town was now almost in complete darkness, except where a light appeared here and there at some window fronting on the street.

The next day Fieldbell ordered a man to roll and empty a few barrels of lime around the well. It was then slaked on the parched soil with water which had also been rolled in another barrel. And, before the Colonel arrived, for he had remained to see the surveyor and his son start together for Mill Creek, the slaked lime had been shoveled down into the bottom of the well to neutralize and render harmless whatever acid had not been burnt off. Before nightfall the pit had been filled and heaped over to wait till the rains should come to pack and level down the earth and obliterate all trace of the mistake.

When George arrived on Mill Creek with the surveyor who had strapped his instruments in front to the pommel of his and George's saddles, they met Andrews hauling his load of oil to the landing. He had, the evening before, reached the works without any accident. The agitator had been unloaded and he had left Mr. Mark preparing to set it up. For this he had engaged the services of a white man who had done a great deal of carpenter work and millwrighting at the Licks, and who had also borne a leading part in the construction, two years before, of the large frame building which covered the retorts. His name was Towle. He lived at a short distance, in one of the hollows upon which he had squatted like the rest of the dwellers in the creek. He cultivated it mainly for corn, potatoes, and turnips, and had built a log cabin in which he was raising a small family of children, the eldest of whom, a son, tended a flock of sheep. From these he clipped his wool, his wife spun it, dyed it, weaved it in her loom, and made all their clothes out of it. They were an industrious family, and he was an intelligent, neat worker.

George introduced the surveyor to Mark. As the survey would occupy the rest of the day, the whole of the next, and most of the third, arrangements were made for the two visitors to sleep in one of the two beds which Mark had bought from Reynolds and for their board at Johnson's, whose wife would them also supply with bed cloths.

Mark went with the surveyor and George to introduce them to Dr. Lewis. Both deeds were carefully read and compared, and, in order to ascertain the exact location of the dividing line between the two thousand acres of the company and the two hundred and fifty acres of the Lewis's property, it was agreed that the meets and bounds of both deeds should be surveyed over again and found as they had been left by to the previous surveyors, the company furnishing one man and Lewis another to carry our chain and drive the pin. The younger Lewis and George accompanied the surveyor who carried and stood up his tripod from point to point up and down the hills through the bushes, trees, brush, briars, and hollows. George had his loaded rifle in his hand and watched upward among the branches for panthers and the like. The young Lewis walked along cautiously also, poking with a stout hickory stick, under logs, dead leaves, and brush for rattle-snakes which the warmth of spring was beginning to bring out of their winter hiding places.

The surveyor pointed to them now and then how the weather defaced blazes cleaved in the bark of the trees by his predecessors, and, when he came to the point of some angle, he would get them to help him search for the old marks on a black oak or some chestnut tree still standing, or on the side of some large stump. In order to obtain the correct place of beginning, he had measured a certain distance down the creek from the white oak between the two entries which the Colonel, two days before, had pointed to Mark and so had found, as the deed named it, a very large boulder imbedded in and rising from the creek, a sharp projecting piece of which had been knocked off for a blaze indication. From there he had chained his line across the creek and across the road, up his side of the hill beyond, and started in search of the points that followed on the deed. At six o'clock they found themselves on the creek again and not far off from the dividing line. Yet as it was getting too dark to obtain a clear sight through his glass, they staked the spot and suspended work until the next morning.

"Well," said Mark, on their return for supper, "you must have gained a good appetite." They could hardly see him through the smoke and coal. Just among the chargers who were all hurriedly engaged screwing up their lids or shoveling up coal into the retorts, while the water out of the pipes was running without stint upon the glowing coke to quench it, spread as it was upon the brick pavement after having been raked down from the retorts.

"Yes," answered George, "the little lunch which we took in our pockets was hardly enough for our long tramp. We will do justice again, I assure you, to Mr. Johnson's supper."

"Did you find all the points, Mr. Lowndes?"

"Yes, so far, so good. After supper I must begin my figuring of the good acres."

After the retorts had been charged and the coke had been wheeled away, Mark joined them at supper, and the incidents of the survey were talked over.

"If the dividing line, Johnson," said the surveyor, "is actually where you showed me on the hill up the creek, the two thousand acres, I think, will be all the ground between it and the place where we stopped this evening, and, from the up lines which we made out today, I cannot see very clearly how there can be two hundred and fifty acres in Lewis's tract down to this dividing line, but this hill is full of cannel coal. Perhaps, farther up the creek, coal is not so high in the vein, and the Lewises, knowing this, may all this time have been trying to secure possession, and possession is nine points of the law, you know. This game is easily played underground where nobody can see it from the outside."

"Then figure up, Mr. Lowndes, after supper, all the land you have gone over.

Tomorrow forenoon, as soon as you have completed up to the line which we claim as our deed specifies its distance very plainly, from the old white oak stump between the two entries, you can finish your calculations for our two thousand acres. After that, you can go on with the shorter survey for Lewis's two hundred and fifty acres from our line to their bounds up the creek. You look very tired however both of you and, if I were you, I would go to bed early tonight to get up early tomorrow morning so as to be refreshed. This would serve you better for these calculations."

They both followed his advice, while Mark betook himself to his books to make all his entries for the day of coal received from every digger and his name, of the car loads emptied by every wheeler and his name, of the bushels of coal charged in the retorts and the name of every charger, of every fireman and all other employees about the works, and also of the quantity and gravity of oil distilled from the retorts and sent to the landing in barrels.

When he had accepted the counterproposal from Col. Albert, he had written to Fieldbell for a set of books to make all these entries. No such books had been kept there before, and the light of his lamp could be seen all night through his windows from the works below by the men on their shifts. Day and night, he was with them, among them, one of them, and as he loved his work, so they loved theirs also and tried hard to get a good word from him about it at all times. Yet, ever watchful as he was, he was careful also to take all the rest which he needed.

At five o'clock the next morning, he opened his door to let the cool air enter the room in which the surveyor and George were asleep. Lowndes opened his eyes and thanked him for the hint. He was soon dressed and at work upon his figures while Mark had gone down to the retorts. George in turn awoke and went down to see how their horses were being fed and taken care of by Andrews and Lewis who had them in charge as well as their own horses. The log stable had been lately undergoing much needed repairs, which Towle, by Mark's directions, had made on the shingle roof, the loft, stalls, racks, trough, and drains.

"Have you watered Selim, Andrews?" inquired George.

"Yes, sir, that's the first thing we done. It don't take us long now for that since Mr. Mark got that trough up below his house to catch up the water from his spring."

"How is the creek water?"

"Getting better every day since the Lewises burn up their tar and their nasty waste from the agitators. It will soon be all right as it was before they came here. Then Mr. Mark says he will have a pipe laid from the big water tanks under the roof over the retorts to fill this new trough here, so that we may always have water here without having to go out for it."

"You cut all your sheaf oats, do you?"

"Yes, sir. This new cutter Mr. Mark also got for us to save feed, makes it better, and saves time too."

"Whose time, Andrews?"

"The horses, sir, as well as ours. It takes them less time to chew."

"What are these wrapt bits for?"

"For next winter, sir. Mr. Mark says we have no right to put bare iron bits in horses' mouths when it freezes hard in winter, to cut the skin off as if they were not iron. He says we have no right to torture these poor brutes that cannot talk, and for fear we forget when winter sets in again, he gave us woolen strips to wrap these bits."

George smiled and laid that up for his father. Meanwhile Lowndes had been busily engaged with his figures and had just completed his calculations as far as he could then when he heard his name called by Johnson from the works below. He looked at his watch and saw it was seven thirty. The retorts had been charged, the coke was wheeled, and the firemen were once more quietly at work tending their fires and cleaning their grates.

"I will go down at once to make arrangements below," said Mark, "so that I can accompany you also as soon as you give me the word." He went to see Towle who had nearly completed the heavy timber frame which was to encase and support the agitator and the iron wheels, and stayed with him until Lowndes came.

"I was right, Mr. Mark," said he, "and I have told it to the Doctor. He is at the line now and will come along with us." Mark followed him and when they reached the spot upon which the Doctor stood waiting with George and the two chain bearers, he said to Mark: "Now, Mr. Mark, if we do not find our two hundred and fifty acres from this line to the bounds above named in the deed, what shall we do?"

"You have your recourse, Doctor, on the parties from whom you bought."

"Perhaps we may have, we will see. Let us go up the creek to our place of beginning as we have it in our deed. Then we will come down to this old poplar which we have also in the deed."

Mark, at breakfast, observing the thoughtfulness of Lowndes, inquired whether his figuring up to his last field notes bore out his surmise that the two thousand acres would most probably call for all the ground yet to be surveyed.

"I think they will," replied Lowndes. "There will be a fuss, I must fear. Have the Lewises hunted much for cannel coal up the creek, Johnson?"

"Yes, they have," replied Johnson, "but the vein there which they have tried in several places is only about thirty inches with a slate roof and a slate floor. Yet here, in this hill, they have with us five to six feet of first-rate bird's eye coal with rock both for floor and roof, very little slate anywhere and but few horse backs."

"Yes," said Mark, "I fear we will have some trouble, unless the terms of their deed actually transfer to Lewis the two hundred and fifty acres within the meter and bounds clearly specified by it."

"I think I will get the Doctor to come along with us and see for himself," said Lowndes. "It is an important matter and he cannot refuse even to mark the dividing line on the creek. I think we can be down here again in time for dinner."

"We will try the Doctor," said Mark.

They began a long march up the hill on the other side of which they crossed the creek to climb another hill till they came to a large rock, a piece of which, resembling the head of a fox, gave them their place of beginning. It was also one of the main bounds specified upon the deed of the company. Having met with the eldest Lewis, they then began a careful search of the points named upon the other deed enclosing the two hundred and fifty acres. At noon having found them all, they reached the foot of the poplar.

"How long will it take you, Mr. Lowndes," inquired the Doctor, "to figure up our tract?"

"About two hours, sir, after dinner."

"Well, I will come down again at three o'clock."

"Very well, Doctor. I will have done at that time. Then I will begin the survey of the mine here."

"Will you survey that too?"

"Yes, sir," said Mark. "Our diggers hear yours somewhere under this line. We want to know exactly how far we can go without touching your coal and how far also you can do the same. I will then get Johnson to sink an air shaft here which will keep us good neighbors." The Doctor gave a faint smile, as if his conscience was not altogether at ease.

"Well," said he, "I will come down at three o'clock after dinner." Mark gave to Johnson the necessary instructions for an immediate sinking of the air shaft upon the dividing line at a sufficient distance from the creek to dig through the coal where the men were at work. As they rose from the table, Mark, Lowndes, Johnson, and George all separated, Lowndes going to Mark's house with all his notes to make his calculations and reckon up the two hundred and fifty acres. At three o'clock Mark and the Doctor came up followed by George and Johnson.

"How do you make it, Mr. Lowndes?" inquired the Doctor.

"Two hundred twenty-five acres and a quarter, sir," answered the surveyor. "This is all I can find."

"All you can find, is it?" said the Doctor, knitting his brow. "We have been working under a serious mistake, and I very much fear, Mr. Mark, that your men who are now digging this shaft will find our diggers beyond the line. The hill has also, I think, the best coal of our purchase."

"It is well then, Doctor, that we should know where we are and act justly toward each other."

"Mr. Lowndes," said the Doctor, "when you return to the Court House, please send to us, from your office, with your bill, your certificate of this survey that I may send a copy of it to the parties from whom we bought this property."

"I will do so, Doctor."

"Now, Mr. Lowndes," said Mark, "please go on with the survey of our mine as far as we have dug. Johnson and George will go with you."

"I will carry the chain," said George, with Ned, in the bowels of that coal. Ned was one of the horsemen who had carried the chain in the forenoon and the day before.

"Let us make the most of the little time we still have this afternoon," said Lowndes, laying his tripod on his shoulder, so that we may finish tomorrow before dinner and be ready to go back at about two o'clock. I will draw the map at home from my notes."

He then went down with George and Johnson. The Doctor had kept his seat after his last words to Lowndes and seemed worried by unwelcome thoughts. Mark stood looking at him.

"I am sorry for you, Doctor," said he, "that this survey turns out as it does. Perhaps if you set an experienced digger at it, he may find out for you that your tract, somewhere up the creek, has more cannel coal in it than you think now, as much coal, it may be, as you have down this way."

"I was thinking of this, Mr. Mark," answered the Doctor, following him out. "We have been so absorbed thus far, these last eighteen months, finishing and starting our works, that we have not had much time for this search, but we will have to make time for it after this."

"As soon as our map is drawn, Doctor, I will show you on it the direction of our vein by the compass, and it may be that, in following it with the compass also on the other side of the creek beyond the hill and the line of our stakes, you may strike it just as we have here."

"This is a good idea, Mr. Mark," replied the Doctor, "and I thank you for it. I see you are going to set up an agitator here?"

"Yes, sir, you remember I intimated as much to you."

"Yes, you did, and we are indebted to you for the idea of burning our tar and our acid refuse. The troughs work very well, and we pile up our coke as you do, using it in our homes instead of coal. I don't know but that we may dispose of large quantities of it at the Licks."

They were thus conversing in the middle of the road leading up the creek when Towle approached to make some inquiry of Mark. Shaking hands with him, the Doctor turned to go back to his works, and Mark went with Towle to the frame which was now ready to receive the agitator. When Lowndes, Johnson, and George came to supper, the length of the main entry, its uptrend, and the rooms opening upon it had been surveyed, the size of each room and of each pillar had been measured, their location had also been numbered to correspond with the stakes intended to be planted on the hillside above, and the same work was laid out for the next morning in the old entry and the branch of the main entry along the creek which had been opened and dug for months upstream to reach the dividing line. While they were at supper, Lowndes looked at Johnson with a smile, and said: "Johnson, some of your diggers have been imposing upon your good nature."

"Yes, I know what you mean, Mr. Lowndes. They have been leaving big pillars and very small rooms to go where the vein is higher and softer for a day's work, but I have changed that as you saw where we went farther in. The pillars are much smaller there and the rooms wide."

"Yes, I saw that. Slaves are but slaves after all, mighty fond of ease and of the crooked ways to get it."

"They have to care however now for the straight line," replied Johnson.

"Yes, the wheelway is not crooked above as it is below."

Mark was listening attentively.

"We will have to reduce the size of them first pillars, Johnson," said he.

"Yes, sir," replied Johnson. "It would be a waste of coal to leave them as they are. I have a good pile of first-rate joints and caps ready for this work. Now, with the starting point which Mr. Lowndes has given us by his compass, we can keep on a straight line from it till we reach the other side of the hill. I am glad Mr. Lowndes came to give us this lift."

"The air hole in that side room which opens on your hollow does not help you much, does it?" inquired Lowndes.

"No, sir, not a great deal. An air shaft is what we need, like the one we have just commenced inside of our dividing line. They will suck up all the foul air from our blasting powder, and we use so much of it lately since we have got into the heart of the bird's eye. It's beautiful, clear of sulfur, and full of oil that coal, but hard like flint to bear under, sharp as our picks may be."

At five o'clock the next morning, they went on with the survey in the old mineshaft, and found it, as they proceeded, still more crooked than the main one which had been opened only a year before. The roof also was slate, and not supported by as many posts as should have been used to render it secure. It had fallen in and filled not a few of the rooms which thus needed badly to be cleaned out before the pillars could be brought down for their coal. As this entry was not so long as the main one, and the rooms were fewer in number, the survey was completed at about breakfast time. Lowndes's report excited again the attention of Mark, while it stirred up in Johnson a good resolution to push up anew this old entry and all the rooms in order to render them permanently safe for further work, even if the vein did not yield quite as well as the new one. It was still mostly a five-foot vein, and the line of the entry was oblique to that of the new one. Between them many new rooms should be dug in.

After breakfast they commenced the survey of the branch which led up along the creek toward the dividing line, and when they reached the last room, they could hear distinctly the hard picking of the diggers who were working for the Lewises. Lowndes, at this sound, stopped suddenly and exclaimed: "They have crossed the line nearly two hundred feet! Go and tell Mr. Mark and Johnson to come," he said hurriedly to the digger who carried the chain with George, both wearing the lard oil miner's lamps stuck in the skull caps they had borrowed. Lowndes and George both sat down upon the pile of coal ready for the wheeler which the man who was working in this room had brought down and shoveled up before they came. They listened attentively to every stroke of the pick just beyond the wall before them.

"It was high time that this survey should be made," muttered Lowndes.

"Yes," responded George, "to stop this kind of work. They will have to pay for all the coal they have wheeled out."

"I have been hearing them come down nearer for more than two months," said the digger, "and I have many times told to Mr. Johnson that I thought they were running on us. He thought so too."

"Do they post well on their side?" inquired Lowndes.

"Well, yes, but they don't share the posts, and this roof is good and solid," answered the digger. Mark and Johnson soon arrived with heads bending down below the low roof.

"How many feet do you make it, Mr. Lowndes," inquired Johnson, "from here to the line?"

"Very nearly two hundred, sir," answered Lowndes. The sound of the pick had stopped for a few moments.

"Now he is putting in his cartridge," said Johnson, "and shortly there will be a blast."

"You have all your measurements on these rooms, Mr. Lowndes?" inquired Mark.

"Yes, sir, and if the men at the shaft outside can make a break through the roof and give us some light, we can go up at once to the Doctor and take him along to survey how far he has come over the line."

Just then a trembling of the bank was felt which was immediately followed by the low detonation of a blast on the other side of the coal wall.

"I left them pounding with a sledge hammer upon a long stout bar to break through the roof," said Johnson, "and it may be that, by this time, they have broken through."

"Let us go and see," said Mark. Johnson's surmise was correct. They were all four getting out of the entry when one of the shaft men came running in from the bottom of the pit, telling him that the bar had fallen through, and was now standing up in the hole, and bearing upon the floor of the entry below. Hastening to the spot, they found the other two diggers standing and waiting for them.

"Mr. Albert," said Mark. "Will you be kind enough to run up to the Doctor and tell him that we would like to have him come down and see this so we may proceed at once to survey his entry on this side of the line. We will let the bar remain as it is till he comes." They sat down, and, in less than half an hour, George returned accompanied by the Doctor and both his brothers. Consternation was depicted upon their faces, and few were the words which the Doctor alone spoke.

"Doctor," said Lowndes, "Mr. Albert and myself would like to return this afternoon, and if you will pilot us into your entry to this bar just where it stands, we can easily make a survey and complete measurement of it and of the rooms from the bar to the place where your men are digging now."

The Doctor looked at his brothers a moment, but soon replied calmly, turning to Mark and the surveyor: "We cannot object to this, Mr. Lowndes, bad as it is for us. We will all go together." Lowndes carrying his instruments and George following him with his chain carrier, they all went up to the upper works. Johnson had instructed the three men at the shaft to remain there and watch for any lifting given from below to bar and for any order which they might receive from him through the hole in the bottom of the pit which was only fifteen feet deep. Should they not however be given any instructions, they were to remain and wait for his return.

On the way to the entry, the Doctor inquired of Mark: "Mr. Mark, how do you think we ought to settle for all this coal dug out of your bank?"

"I do not think our company, Doctor, are disposed, under the circumstances, to be hard on you. As far as I am concerned, I would regard the value of your posting and of the wheelway to be about equivalent to the value of the coal carried out. Still we need this survey to complete our map to this day."

The three brothers heard Mark's reply and their manner showed evident signs of relief. When they reached the wheelway, they entered into the mine, and, soon, by the light of the little tin lamps, discovered the pointed end of the long heavy bar resting upon one of the tiers under the rails.

"Your guess was a good one, Johnson," said the Doctor, "for the location of this shaft."

"Oh! I took my bearings from our main entry, Doctor, and then from the edge of the bank along the creek."

"Johnson has done very well with his eye and his head," said Lowndes, "but I have given him points by which, with the stakes outside which I will plant on the hill, he will be able always to tell the rooms here below when he walks above over them in broad day light." Then he went on with his measurements from room to room, noting down the size of each and of the pillars which separated them, and numbering them all till he had reached the last one. The digger was there whose pick strokes and blasting they had heard so distinctly on the other side of the wall.

"We have to get out of here, Jack," said the Doctor. "This coal don't belong to us from here to James's room above."

"Is that so, Doctor?" said Jack. "Well, many a time I have thought, from the way it sounds on the other side when they dig and they blast, we must be very near, very near the line."

"We have passed it, Jack," said the Doctor.

"Of course we must get out and leave this pile here too." Gathering his tools, he sat down on his coal to wait for the wagon, in which he would load them to be carried to another room above the line. On their return to the shaft upon the hill, Johnson directed the three men to enlarge the opening in order to crib it up with large hewn timbers which he told them to go down and bring up at once. But he would first however crib the roof at the bottom himself and make it secure with strong posting.

With an axe then he cut down two young saplings from which he proceeded to split and joint stakes for Lowndes to mark upon the hill the line and the direction of each entry, and the mouth, length, and width of each room in each entry, the size of each pillar being thus also outlined. Two hours were occupied after dinner for the completion of this work by the aid of Lowndes's notes and instruments, after which he and George mounted their horses and started on their return. As they left, Mark handed to George for his father a full account of the settlement with their neighbors above. He also informed him in the letter that the agitator was now set up and connected with the tanks, and that, on Monday following, the charging of the retorts would be changed from every twelve hours to every eight hours of the six days, all the men having readily agreed to the additional labor which would entitle them to receive each personally ten dollars every month in cash. They would make the first charge in the morning at four o'clock and the last at eight in the evening.

The wood firming was also giving good results. With the leather worked well and with the caustic potash which they had already begun to lay up, they would start the agitator the next day. They could then make use of the additional tar troughs which they had already prepared, and would have more coke to sell about the Court House or wherever a paying market would be found for it down the Kanawha.

This letter reached the Colonel in the evening a few moments before the supper bell had been rung. A broad smile of satisfaction lighted his face while he was reading it in his room to his wife and his son.

"Well, George," said he, after he had read the last line, "you and Mr. Lowndes have accomplished a very great deal in a very short time, and Mr. Mark deserves a great deal of credit for piloting all these things, especially for keeping the peace between Lewis and our company. I certainly think as he does about the value of the entry, as it now stands, being equal to the value of the coal which has been dug out. Johnson can go on now making permanent arrangements for his entries and his rooms since the dividing line is well established between the two properties. The old entry will now receive his attention and our coal there will be saved from waste. Then think of the work of his agitator. How quickly also he fixed it up! I must read this letter to Fieldbell after supper. It will or should spur him to do likewise in his department."

An hour later however when the Colonel for the second time read this letter in the same room but in the hearing only of Fieldbell, he could see that the refiner was not as much pleased as he himself.

"Will you not get into trouble with the owner, Colonel," said he slowly, "by paying ten dollars every month to each of these Black hands?"

"Had you any such trouble yourself, sir, when you paid Reynolds's orders for their extras?" the Colonel inquired, looking fixedly and severely at him. "These extras were for digging coal which came out of the retorts almost as green as it had been charged and is now heaped up in ashes along the creek, but these extras are for charging coal, every lump of which will give us all the oil it holds and for adding one third to our production. As for the owner, we are not here in South Carolina nor in Mississippi, but in a border state and very near the Ohio River. They know it too well to say one word against it. I am a slave holder myself, on a small scale however, for I have been very careful, these last twenty years, to reduce this kind of stock as fast as I could honorably do it. It's an infernal stock. But, leaving that aside, it is just that these men should be paid for more work than what had been named to their owner. Our duty will be to take still greater care of their health and of their general well-being for the increase which their work will make in our profits. I know, for me, that Mr. Mark will pay special attention to this."

"Oh, mind, Colonel," said Fieldbell, when he saw the honest indignation which he had excited by his silly remark, "I do not take the part of the owners."

"I don't suppose you would, Mr. Fieldbell. You have not been trained in that school, but I have, and I know, from bitter experience, what seeds of barbarism it has for sixty years been sowing broadcast over our South. I don't wonder at the nomination of Abraham Lincoln yesterday at Chicago. I don't go the whole length he does, but I am more than halfway with him. In the end however, as he well says, freedom and slavery cannot live together side by side in this fine country of ours. Slavery must go to the wall by fair means if it yields, by force if it won't!"

Fieldbell was a Northern man. He nodded in assent, and the enthusiasm of the Colonel had a magnetism in it which, for the moment, caused him almost to feel shame for the utterance of his suggestion.

"I am glad," he felt constrained to say, "that Mr. Mark is doing so well. This mild alkaline purging at Mill Creek will give us a fine burning oil out of the stills here."

"Won't it lessen your work too, even if it increases our product?"

"I think it will, Colonel. The barreling only will be increased."

At last, thought the Colonel, I have brought you to own it. They convened a while longer, until Mrs. Albert came up, when Fieldbell retired to his room.

"I have read this letter to him," said the Colonel to his wife, "and it is wonderful what a soothing effect the honest course of Mark is having upon him, knowing too that we only pay to Mark half the salary which he is getting himself."

"Mark must indeed be a good husband," replied Mrs. Albert, "and his wife must be like him and a tender loving mother. As soon as she can bear the journey, arrange it so that there shall be no delay in their coming together again."

"Yes, I will keep on this lookout. I have told him already," answered the Colonel.

After breakfast, the next morning, he called on the surveyor at his office in the Court House and found him at work upon all the maps of the mine, its entries, its rooms, and the configuration of the hill above it. The measurements were all there and the crooked line of the wheelways.

"It was high time, Mr. Lowndes, that you and your compass went there together to stop this wild sort of work on that rich mine, was it not?"

"It was indeed, Colonel, but you have a good man there in Mr. Mark, and another honest worker in Johnson. He is so well pleased himself now to have points and lines to start from and go by. This whole survey has been very pleasant to me. Lewis felt pretty sure, but I think, on the whole, they will gain a better acquaintance with their coal lay by the operation."

"Yes, Mr. Mark wrote to me that he instead showing your map to them that they may see where they had better dig to find on their side of the creek the continuation of our vein in the hill."

"That's it, that's what I mean. They don't know actually yet what they have, because they don't know where to look for it to find it."

"How soon can you let me have this map?"

"It will take me about three days to make it and also the map of your two thousand acres."

"Well, go on with your work, I won't disturb you. Bring the deed along when you come to me, also your bill. Good morning."

Chapter Seven

Husband, Wife, and Children Together Again

Three months have passed. Mark has succeeded in establishing, by actual results, the correctness of all his figures which he had given to the Colonel. Fieldbell acknowledged that every one of the changes worked out by Mark are valuable improvements for which the company had but done justice to itself by raising his salary to the figure of his own without waiting until the expiration of the six months. The old quarters of the Blacks employed at Mill Creek have also been enlarged and much improved, and their daily comforts are being generously provided for in return for their growing attachment to the place, and the lessening of their old weekly hankering for their wonted but wearying foot journeys to and from town. The services of a worthy colored Baptist minister have been secured to preach the Gospel to them on every Sabbath when Mr. Freenberry did not preach at the landing, and the orderly deportment of the men during the week at their work and off their work is an encouraging indication that these Sabbath ministrations are appreciated and bearing good fruit.

Mark's wife and their four children have been with him in their new home for two weeks. Col. Albert and his wife had brought them and their baggage in a large spring wagon in which they had returned to their own new home also the same evening, both ladies highly pleased with each other. Samuel was the eldest child, nearly nine years old, Harriet was his sister, Nathan was the name of their young little brother, and Sophie, the youngest of all, was hardly two months old, a quiet, pleasant, black eyed little girl, whose round little face so captured the heart of Maria, the good natured old cook, that she declared there was not another such pretty child in Kanawha county.

Many a time, during the day, she would steal away from her work to come up and kneel down by her cradle to take a long look at her, speaking low words of love, and begging the favor to rock her to sleep whenever she saw, with the mother, that the child needed it. At her own suggestion, her oldest unmarried daughter had been engaged by Mark at the Licks from Mrs. Wright, to assist his wife in the kitchen and in the house. To the latter another log room had been added by special request from Mrs. Albert, who, while it was being constructed, had come up several times on Selim the horse to satisfy herself that it would be in perfect order before Mrs. Mark would arrive with her children. The kitchen had also been framed larger to make a recess room for the bed of the girl whose name was Alice, and whose natural dispositions were much like her mother's: quick and neat in her ways, and fond of children. She had a high temper, was easily provoked, but appreciated keenly kind and gentle treatment, and responded to it with intelligent obedience, ever ready even to anticipate unspoken wishes from Mark or his wife.

Toward both and their children she felt more and more attached every day, sitting with them at the hours of worship and sometimes bringing her mother, especially evenings. At these times, her father also, old Tom, if he was not at work below, would, as he knew well the hour, make his way up the rear of the hill, unobserved, and through the cracks of the back plank door, near the kitchen, listen with all attention to the reading of the word, the prayer of the father, and the hymns sung in unison with the mother's accompaniment upon her melodeon. At such times of family communion with God, in the morning at six o'clock, after the first charging, and in the evening at the same hour, before the last charge of the day, the sacred melody softly descended the hill and was caught by other listeners in the cabins below, teaching in peace welcome tunes to each and good will to them all.

Mrs. Johnson and her daughters had prepared the noon meal at Mark's house before he arrived with his wife and children in the Colonel's conveyance, and to Alice, who had also come in that morning on a horse from the Licks, they had committed the care of "Lucky," the cow so highly recommended by Mrs. Reynolds. Brown robed and white spotted, she was a harmless, quiet animal, always telling where she was browsing by a little brass bell strapped to her neck. As Selim was as gentle as she was, they from that day became almost inseparable, so that when Alice and the children, mornings and evenings, started out to hunt her for her milk, and hearing her bell through the woods, would call her by name she would soon make her appearance to get her warm mush in exchange for her milk, followed at a short distance by Selim who would then and there quietly wait until she was free. Then he would follow the three down to the rail fence in the rear of the home, where he would stand till his ear of corn had also been brought to the ground before him. While he contentedly crouched there, the chickens who had followed the children all found in him a liberal, generous friend ready to share with them.

After breakfast, Samuel and Harriet, in their mother's room, would betake themselves to their school books and memorize their lessons to recite them to her when she would call for them. She would then add her comments and explanations and increase their interest in the information thus daily acquired. Mark, at dinner, would also question them upon the same studies and add his own stores of knowledge to those of his wife so that both children, under this wise mental training and open air exercise, were finding in their new and wild surroundings, if not the sharp teachings of the crowded city school room, a salutary exemption at least from its multiplied temptations and dangers.

Mark was, at dinner one day, explaining to them the advantages of railroads over the primitive wagon roads.

"When are you going to begin building your tramroad to the landing, Papa?" inquired Samuel.

"We don't know yet, my son. It will cost a great deal of money, and, before we begin spending it, we want to know for certain whether we will not very soon be obliged to sell our oil much cheaper than we do at present."

"Why?"

"On account of the great quantity of oil which people are now every day in Pennsylvania getting out of their oil wells without even having to pump for it."

"They don't pump for it?"

"No, it flows up in many places where they bore for it just as you saw the other day, or like Mr. Wright at the Licks, these men on the river bank boring for salt water, from a high derrick and through an iron pipe. They are beginning to refine that oil, and they can afford to sell it very cheap."

"But you say it does not last as long as ours does in the lamps?" inquired his wife.

"No, not so long, and it explodes easily while ours does not. Still they are improving it all along, and, because it is cheap, people are beginning to use it very much, people who have been using ours till lately, so that we are compelled every week to make our prices lower in order to keep our customers. Another thing, we don't know but that we may have war here before long."

"War, Papa?"

"Yes, Samuel, and a dreadful war too."

"What war?"

"You see these Black men down in the works?"

"Yes."

"Well, they are slaves. They were born slaves, not free as we are. They do not belong to themselves but to rich white men, from whom we hire them, and who want us, white men too, who won't own slaves in the states north of the Ohio River, to say that we are willing to let them bring their slaves among us in the Northern states and make them work upon farms as they do in the Southern states, but we won't allow it. And because we tell them to keep their slaves at home, they are very angry, and they declare that they will force us to consent to it by making war on us."

"Would they come with guns here and cannons, Papa?" inquired Harriet with terror in her face.

"Perhaps they might, but we won't let them come here. We can get plenty of guns and cannons too, and their slaves here would help us pretty quick."

"That they would," cried Alice from the kitchen. Ever on the alert, she had been listening as attentively as the two children and had heard every word. Mark and his wife smiled while looking at each other.

"Oh, but I hope," continued Mark, "that they will have sense enough to let the guns alone and listen to reason. If they do so and we remain sure that they will make no trouble, we will go on with our tramroad, unless however that Pennsylvania oil gets to be so plenty and so cheap that we may not be able to sell ours as cheap as it. So, you see, Samuel, we have to wait patiently before spending so much money on a new road."

Such was also the mind of Col. Albert. They had both lately held anxious consultations on these ominous signs of the times. They were nearing the close of August. On the 4th of November, a successor would be elected to occupy in March following the place of President Buchanan, whose vacillating mind had sadly proved unable to master and control the bitterness and antagonism of the four parties surging around him. If Abraham Lincoln should be elected over John Bell of Tennessee, the nominee of the Baltimore Union Constitutional Convention, which had been held the 19th of May previous, also over Stephen Douglass of Illinois, the nominee of the Northern Democrats, and lastly over John Breckinridge of Kentucky, the Charleston nominee in April of the Southern Democrats who were openly advocating the reopening of the African slave trade, would not the immediate issue be the disruption of the Union as threatened by the South, and what then could or would Abraham Lincoln do in such an emergency?

What would the Northern states do also, divided as they were among themselves? There were overshadowing questions. The startling discoveries of petroleum in Pennsylvania puzzled them also, but so far none of that petroleum had been found to contain the lubricating qualities of the cannel coal oil, and as long as this advantage remained, the changes introduced by Mark had more than enabled the company to lower their prices on the burning oil in order to hold their own against that extracted, distilled, and refined from crude petroleum.

The Blacks in the works were also keenly on the watch for the coming events, and daily communicating to each other any and every bit of information overheard from the whites relating to the extinction or the prolongation of slavery. None of them in the works could read, hence the Cincinnati papers were, to their great sorrow, a dead letter, a foreign tongue to them all.

Yet Samuel was a favorite among them. He had taken a fancy to Pole the singer, and Pole had become his scholar. The days were getting a little longer, and, after supper, when coal wheeling out of the bank was over, the side pit from which Nagg dug his fire clay for the retort lids to the rear of Alice's kitchen became a sort of animated school place where, as long as the light of day should last, Pole could be found kneeling earnestly on the ground against a log on which Samuel sat teaching him to spell from one of his old school books from Chicago and Newark. Pole could at any time, when in town, for a dime, buy one of those Cincinnati papers, and the great ambition had lately sprung within him to learn to read, so that the paper could remain no longer sealed against him and his friends. Samuel was joyful that his own progress could already qualify him to teach another so much his senior in years. He had made his mother his confidante the first evening when he came on a run for his spelling book at the urgent request of Pole.

His mother had given her consent, provided Pole would give the promise that, as soon as he could read, he would buy a Bible printed in large letters, and Pole had gladly made the promise to Samuel. Alice too had of course enlisted the ready sympathy of Harriet, and the race was now between brother and sister whose scholar would first be able to read in the Bible, for Alice cared little for the Cincinnati paper. She was eager to be free someday, and she felt instinctively that the reading of the word of God would make her free. Her mother and her father had always quoted much of the Bible to her from memory, for if they could not read, they were good listeners and could remember all that was read and taught to them out of the Bible. It is well to add here that Mrs. Mark had told to Samuel and Harriet in the first place that the best disposal they could make of their old spelling books was to let Pole and Alice keep them to memorize their lessons between working hours, and thus prepare themselves for reciting hours. In their joy at the gift, both scholars had obtained from their young teacher the inscription, by their own hands, of Napoleon Gibbs inside the cover of one book and of Alice Wright inside the cover of the other. With their own names thus attached, they felt the more certain soon to be able to read them without trouble from beginning to end.

Pole however made more rapid progress than Alice. Wherever he could squat or sit down, he would take his book out of the breast pocket which he had sewed for it inside of his flannel shirt, and begin to study it for a while, either in some room in the bank by the flickering light of his little tin lamp, or in their cabin. He was then at once surrounded by other Black learners as anxious as himself to make the printed letters and the printed words their own by dint of hard work. Teaching them he would teach himself, and would surprise Samuel by spelling and reading sometimes pages beyond the lessons which he had given him to recite at their next session in the clay pit.

When November arrived both Bibles had been bought at the Licks, and this time Mrs. Mark had been formally requested by Pole and by Alice to write their names, in each, on the first fly leaf, and also the day on which they had become the owners of the precious volume, after which Alice had carefully sewed black cloth around the covers to keep them clean from daily use. Pole had bought them both with his own and Alice's money (she sewed often for the men), and, with the help of the colored minister, he had been careful to choose a clear large print which they could easily read. The store keeper at the Licks who had ordered them for him from Cincinnati shortly after had to order several more to supply other demands from Pole's school mates, and some of them by this time would now and then, in a way of their own, get hold of a Cincinnati paper which they would bring to Mill Creek to decipher the news in it at night among themselves, Pole being the best able to do so.

Well they might, for momentous events were rapidly succeeding each other. The 4th of November had passed. Abraham Lincoln had been elected President of the United States over the three other candidates. Governors of the slave-holding states and the leading Southern politicians openly declared themselves in favor of secession as the only means to avert the calamities which a Republican administration, they declared, would surely precipitate upon them in proclaiming the immediate abolition of slavery, and in inciting the negroes to rise, as John Brown had done. Both houses of Congress had appointed committees to try and harmonize the antagonistic elements. These committees had failed, and on the 8th of November, the Palmetto flag had been hoisted in Charleston, and the two United States senators from South Carolina had resigned and returned home. South Carolina had voted $100,000 to purchase arms for the state;

Louisiana $500,000. Georgia $100,000. On the 3rd of December, Congress had met. President Buchanan, in his message, had taken the position that no state had a right to withdraw from the Union, but that the general Government had no authority to coerce her if she did.

Against this piping utterance a ringing protest had been heard in response from far off Tennessee over the grave of Andrew Jackson, and, in similar disgust, on the 14th, General Cass, Secretary of State, had resigned and gone back to Detroit. On the 20th, South Carolina had seceded by the unanimous vote of her Convention at Charleston, and, upon the reception of this news, on the 21st, at New Orleans, one hundred guns had been fired and the Pelican flag had been unfurled. On the 26th, Major Anderson of the United States Army had evacuated Port Moultrie after spilling his guns and destroying the carriages and had occupied Fort Sumpter during the night. On the same day, in the South Carolina Convention, a resolution had been offered looking toward a Southern Confederacy, and on the 24th, the Governor of South Carolina had declared that state "separate, sovereign, free, and independent, vested with all the rights that appertain to a free and independent nation." On the 1st of January 1861, preparations were being made to take Fort Sumter.

"What are we coming to, Mr. Mark?" anxiously inquired Johnson. "Are not these people getting crazier and madder every day?"

"They are, Johnson. Bent upon burning the roof over their heads. Here is a letter received in Washington from Alabama, which declares that, if a single gun is fired in opposition to disunion, Mr. Lincoln's life is not worth a week's purchase. They have besieged Fort Sumter, cut off all communications, and everything is in readiness to open fire on the Fort."

"If that fire is opened," said the Colonel who came up with his son a few days after, "Major Anderson, who will soon be out of provisions, will respond to it at once, the Kentuckian that he is. Then war will begin, that civil war which we have all along dreaded. The North will start up and stand like one man, and all these Black men will be off in Ohio, free like the air which we breathe."

"Not all, Colonel. A few may, a very few if their owners, at Christmas as you saw plainly, were but too glad that we consented to keep them without any charge all this year, if we would board and clothe them. These men appreciate the way we treat them. I hardly think any of them will leave us. Home is a dear place to every Negro. They cling to it till forced away by bad treatment, and the North, to most of them, is a foreign country which they have never seen and never tried."

"That's true, Mr. Mark. This is their character, but what of Virginia?"

"As Richmond goes, she will follow, I presume, and Richmond is in very bad hands," answered Mark, "but this Kanawha country, Colonel, is not like Virginia east of the mountains. It is not agricultural. Her wealth consists in coal, iron, and salt. The Blacks here are few and far between. The whites predominate greatly in number, and these whites, if poor, are industrious and make good wages. They have most of them come over to put the mountains between them and the Blacks and their owners. This country will stand with the North and will be glad to get rid of the few rabid politicians from the East who leave no stone unturned to force Virginia out of the Union."

"Yes, but those politicians own all the property here. It is their money which digs the coal and boils down the salt."

"Very well, but you know their men are mostly whites, not the white trash of Georgia and South Carolina, but live men like Johnson and Druker, and Towle, who will give all aid and comfort to the North in a struggle of this kind."

"You have no fear then for the safety of these works and of our refinery at the river?"

"The salt works, Colonel, are a necessity for all the towns on the Ohio and the upper Mississippi. The North won't give them up. They will be protected, and the same protection will cover us. Our own firm and open stand here for the Union will encourage the other oil companies and will also be joy and strength to all these mountaineers around us. It will nerve their hands for the right when the trial comes."

"Sound reasoning this, Mr. Mark. I have no fear for this property with you upon it, but I wish I could say as much of Fieldbell and of our refinery, for, to speak to you candidly, I am thinking seriously of returning to Kentucky with my wife and my son. I have been written to these last few weeks repeatedly from Frankfort, and I am urgently wanted there, they say, to strengthen, as you do here, the hands and my heart of our loyal friends. Conspirators are at work plotting to drag Kentucky out of the Union."

Mark looked at him inquiringly, and the Colonel continued, "Fieldbell is anxious about the storm which is coming, and not anxious only, but afraid also. From a few intimations of his, I and my wife see pretty clearly that he would be glad to return to Ohio."

"Out of range?" said Mark with a smile.

"Yes, long range or short range, any range. Could you handle both works, Mr. Mark?"

Before answering, Mark reflected a moment. Then speaking slowly, he replied: "Colonel, the refinery ought to be here as Lewis and others have it. The move would concentrate labor and eliminate much of it which is now very expensive. With one half of the fixtures brought up here, I would easily refine all our present crude product. The other half could remain there for the present. And, while I am seeing to their transportation and setting up, Johnson would assist me very efficiently. He is intelligent and reliable, and the change would enthuse him as well as all these men."

As he spoke, the Colonel listened very attentively, while his eyes sparkled with content.

"As I have told you before, Mr. Mark, you have indeed a sharp eye. And this idea is a great relief to me. What salary should be given you?"

"Oh, well, I leave that with you and Col. Truesdell. This measure is one of safety, but first of economy. The books will show you the saving."

"Fieldbell is now President."

"As for this, Colonel, excuse me for interrupting you. I have a condition to submit, and it is that you will consent yourself to be President of the company, my part being that of superintendent of the works."

"Well, well, we can arrange that in Cincinnati as soon as I get back," said the Colonel. "Let us now go up to the house. I want to see your wife and bid good-bye to her and the little ones. My wife had thought of coming with me instead of George, but we did not know whether you would see your way clear to accept my proposals. I feel very grateful, I assure you, for your clever turn on it, but I will not speak of this to Mrs. Mark. You will explain it all to her after we are gone. Tomorrow or day after, I will write to you my settlement with Fieldbell and will enclose in my letter by Druker all the authority which you will need for managing and concentrating both our works."

Thus did the two friends part on the eve of the coming storm, the employer and the employee, both actuated by the same spirit of loyalty to their divine master and to each other as His servants and followers, honoring His command to bear one another's burden, and so fulfilling His law.

As January wore on, Fort Pulaski, at Savannah, Georgia, was taken possession of by the Georgia state troops. Delaware, by a vote of both houses, refused to secede in accordance with the wishes of Mississippi. The arsenal at Mobile, with all its vast stores of powder and ammunitions, was taken by the Secessionists, and Fort Morgan, at the entrance of Mobile Bay, was, in the same day, occupied by Alabama troops. The Governor of Virginia, in his message to his Legislature, declared that he would regard the attempt of Federal troops to pass across Virginia, as an act of invasion which must be repelled. On the 9th, Mississippi seceded. On the same day the steamer *Star of the West*, sent from New York by order of the President, in her attempt to enter the harbor of Charleston, to provision Sumter, was fired upon by the Rebel battery on Morris Island and forced to put to sea again. Governor Pickens, answering to the demand of Major Anderson for explanation, replied that this repulse was in accordance with his orders. On the 11th, Florida and Alabama seceded. On the 12th, artillery was posted at Vicksburg to hail boats passing on the Mississippi. On the 19th, Georgia seceded by 208 to 69. On the 20th, Jefferson Davis, senator from Mississippi, withdrew from the senate. On the 26th, Louisiana seceded. On the 31st, the United States Branch mint and custom house at New Orleans were seized by the state authorities and $571,000 in Government funds were captured. On the 8th of February, in the Montgomery Convention, Jefferson Davis was elected Provisional President and Alexander H. Stephens Provisional Vice President of the Confederate States, and the Little Rock arsenal was surrendered to the State of Arkansas with nine thousand stands of arms, forty cannons, and a large amount of ammunition.

On the 13th, the Electoral votes were counted in the National Congress, and the Vice President Breckinridge declared Abraham Lincoln, of Illinois, elected President, and Hannibal Hamlin, of Maine, Vice President of the United States for four years, from the 4th of March 1861. On the 23d, Abraham Lincoln arrived in Washington, having secretly left Harrisburg during the night to avoid a threatened assassination. On the 1st of March, Major Anderson, at Sumter, reported the works surrounding the Fort nearly completed, and that further delay was almost impossible. On the 4th, Abraham Lincoln was inaugurated President of the United States, and Texas seceded. On the 5th, General Beauregard was ordered by President Davis to take command of the forces for the investment of Fort Sumter. On the 22d, a special messenger from Washington, Dr. Fox, visited Fort Sumter. On the 4th of April, the Virginia Convention rejected, by a vote of 89 to 45, a Resolution for an Ordinance of Secession to be submitted to the people.

During these three months of stupendous folly in the South, Mark, buoyed up by an abiding faith in the justice of an overruling Providence, had steadily, day after day, been moving equipment up from the old refinery to the new one, such as the large stills, agitators, and tanks as he needed, and had reset them under a new building separate from the old, thus closing two sides of the flat formed by the bend of the creek. He had so arranged his movements however that no interruption had been permitted to occur in the work of the retorts, nor in that of the refinery, nor in the shipments to Cincinnati, and the letters which he received every few days from the Colonel encouraged him by their manifestations of surprise at his being able, through such a removal, to keep the company so well supplied that they could meet a constant demand incited by the superior quality of his products.

This change had also rendered necessary that the shipments of the refined oil from Mill Creek to Cincinnati should be made in new barrels, and the first shipment had been carried by Druker to the steamboat at the Court House. On his return to the landing, he hastened to bring to Mark the information that the old Governor, General Henry Wise, had just arrived from east Virginia with two companies of the Richmond Greys commanded by his two sons, the same Henry Wise who, eighteen months before, had hanged John Brown at Harper's Ferry.

"He comes here, Mr. Mark, to raise an army from amongst us, the Kanawha boys, he calls us, to hold the salt works and the river against an invasion from Ohio as he calls it too, but he will have to whistle a long spell for us to make up his army."

Johnson, Towle, and three other white carpenters were with Mark at the time and looked at him at once. He stood for a moment stunned by the news. But, soon recovering himself, he replied: "His old Richmond friends at the Court House have doubtless written to him to come on this wild goose chase and have made him believe that his name and his great popularity would in a short time rally around him as many regiments as he would want for that grand purpose."

"The old fool," exclaimed Druker. "He don't know us yet with all his knowing. The Kanawha is not Richmond. Let him try and send his Grey chaps or green horns up the Elk, and if they don't learn to throw their guns and run like hounds before they have made four mules, my name is not Druker. The idea to drag us into their mess! Do they think that we are born fools as they are!"

"What about the steamboat?" inquired Johnson.

"The captain had heard in the early morning of their coming down from Hawk's Nest and kept on the lookout. As soon as he heard their drums and saw them on the march from the Licks, he blew the whistle, pulled his planks, and was way down the river before they reached the landing. He was up to that game. They were too late to nab him and his boat."

"Yes," said Mark, "everything South of the Ohio they think belongs to them."

"They will soon find their mistake, cried Druker, "and be glad to let go."

By this time all the Blacks had come from every part of the works, surrounding Mark, Johnson, and Druker. They listened eagerly to every word they uttered. Mark's serious countenance reflected the conflicting thoughts which were rapidly passing through his mind.

"Druker," said he, "we must open their eyes."

"How, sir? Give the word."

"As you go home, leave word and have it pass around all along the creek and up the Elk, that, beginning tomorrow, all Union men will meet every day at two o'clock in the afternoon in the large old field opposite the landing, to sign our names to a paper to be sent to the Convention at Richmond protesting against their vote of the 4th of this month which refuses to submit to the people of the state their ordinance of secession, and to declare our resolve never to bear arms against the United States of America."

"That's the move, sir. I will start it right." And, in his eagerness, he turned to go.

"One word more, Druker. We will have to stop boating and shipping for the present till Wise and his friends go home and leave us quiet. They have, I suppose, taken possession of the post office, and we cannot send any letters for the present to Cincinnati nor hear from there, but that cannot last long. Let us take our stand at once, however, to let them know that they cannot manipulate at their own sweet will. I will go and ask Lewis to join us. Charley is outspoken and will not be afraid to come out, but the Doctor and Alex may not want to commit themselves, Union men though they are. At any rate, they will have soon to show their hands, whether they will or not. They cannot sit long on that fence and look on with their mouths shut."

"Let them alone, Mr. Mark. If they begin now to make faces, we don't want chicken-hearted fellows with us," exclaimed Johnson. "Go ahead, Druker, and lose no time, that we may be as many there tomorrow afternoon as possible."

"And make a demonstration," added Mark, "which will speak for itself at once down at the Court House and set Wise and his folks a-thinking that all is not roses before them."

"Hurrah! That's the talk, Mr. Mark." And with hasty steps down the creek, Druker was soon out of sight.

The Blacks, leaning upon their shovels and their picks, still kept around Mark and Johnson, glancing at each other, not afraid, but grinning and wondering.

"Are they going to take us to Richmond, Mr. Mark?" asked old Tom.

"Not a bit of it, Tom. We want to show them tomorrow and day after, as long as they stay down there, that they had better not disturb us among our hills here. Don't be afraid, any of you. We are more than they are, and moreover we have the right on our side. Go to work and don't mind them. They will soon be back where they come from."

They knew well the man who spoke to them so resolutely; at his word they quietly went back each man to his work.

"Johnson," said he, at the same time addressing the four white carpenters, "we will arrange to go down together tomorrow to that rally."

"Yes, sir," all answered.

"We will take our rifles along," said Towle.

"Yes, no objection to that. We might have told this to Druker but it is not too late yet."

"We will go up the creek," said Towle, "and send word along the hollows."

"Very well, do so. We must have a good representation from here."

Before leaving, they all went to box up their tools, and he went up to tell the news to his wife. She had seen from the house that something uncommon had happened below in the works and was somewhat anxious to learn what the cause of the stir could be, especially when she and Alice observed Druker hurrying down the creek.

"What can be the matter?" she inquired of her husband as soon as he came up.

"Druker says that Henry Wise, with two of his sons and two companies of Richmond Grey as they call themselves, have just arrived at the Court House to take possession of the salt works and mean to close up the Kanawha against all trade with the Ohio."

"What a foolish old man to come so far from his home on such an errand!" she exclaimed.

"The barbarism of slavery, my dear, is the root of all this, but their bluster will recur on themselves. I have just suggested to Druker, Johnson, Towle and his three men the idea of a general muster of all our Union men to meet every afternoon beginning tomorrow in the large field opposite the landing, to sign our names to a protest to be sent to the Convention at Richmond against their refusal to submit to a vote of the people of this state their ordinance of secession. We will also declare in the same memorial our resolve never to bear arms against the United States of America. I must prepare this paper."

As he sat down at his desk, his wife, standing by him, said to Alice who had come from her kitchen to the back door: "You was right, Alice."

"How so?" inquired Mark, looking up.

"She was telling me, just before you came up, and as we were looking down upon you all, that she had heard her father speak of Wise and his sons and of these two companies coming down from Richmond."

Alice smiled in answer to Mark's inquiring look. "Yes, sir, we all know what they are doing better than you white folks do. We knew they were coming about a week ago."

"Yes," replied Mark reflecting. "The thing must have leaked out through his friends at the Court House. They had correspondence with him." He went on with his writing, while his wife who had followed Alice into the kitchen, was answering as well as she could the many inquiries of Samuel and Harriet about General Wise and his gray soldiers, the muster on the Elk, and the war. His patriotism however did not hinder Mark from attending scrupulously to his duties in every part of the works. A special one was laid upon him by the unexpected arrival of Wise. He did not know how long he would be permitted to bar the river to Cincinnati, but in the meantime there must be no stoppage in the production of the works, and the new barrels filled every day must be laid in safe places till the embargo could be raised. After having written the memorial and signed it, he went up to Lewis to propose it also to them for their signatures.

"Ought we not be very cautious about this matter, Mr. Mark?" answered the Doctor. "Oil properties can easily be set on fire by desperate men, and your responsibility is great, Mr. Mark. Do you not endanger, by this step, a large amount of property which is in your hands only for safe keeping?"

"Do I not ensure its safety, sir, by raising a large force here of willing and determined men to protect it from being seized and carried off by their wagons while they can do it? Are you going to wait and tamely submit to this wholesale robbery? No, sir, let them understand that we are ten armed men to their one in these mountains and that they had better stay down there and fill their wagons with salt if they are let to do so, than come up here with them to be shot at from every hillside on the way. I am sorry for the oil works on the river. I suppose they have all stopped and have shipped off all they could before Wise made his appearance. But we are safe here and can only assure our safety by showing to him that we understand thoroughly well the strength of our position. He cannot stay long in this valley. He will soon find out how he has been deceived into a trap by foolish friends when he is driven out beyond Hawk's Nest. Ours of course will be a ready market in Cincinnati for all our accumulated stock."

"Mr. Mark is right, Doc," exclaimed Charley, the younger brother, while both his elders looked at each other smiling their assent to the idea of a fine market in prospects.

"It is true, Mr. Mark," said the Doctor. "He cannot stay long here, and he would not dare to come up this way, but these Secessionists are desperate. One or two of them, in revenge, might be easily induced to steal out at night and set us on fire."

"Our Blacks are up to that game, Doctor, and they will scent it so effectually, when warned of it, that day and night they will be on the lookout and watch for newcomers who they don't know to be Unioners to the core."

"Well, we will sign, Mr. Mark." The three of them wrote their names below Mark's. "But we cannot all be down at the landing tomorrow," said the Doctor.

"I will be there," said Charley, "with all the white men that we can spare, with our rifles too."

"Very well, this is a good beginning," said Mark as he left them.

The wife of the Doctor was in the next room during the conversation every word of which she had listened to and heard. She had often seen Mark and thought highly of his character. "Mr. Mark is certainly correct in all he told you, Doctor," she said. "His motives have the right ring, and the course which he counsels is doubtless the safest because it is the wisest."

"I don't say no, wife," replied the Doctor. "But these are dangerous times, and we have to be very cautious. Mr. Mark's reasoning however is correct and sound."

"And that is a true Union man," she added.

Chapter Eight

Loyalty Tested

The muster gathering the next day was very large and enthusiastic and, when upon the suggestion of Mark, a long line of all the men present was formed inside the rail fence on two sides of the old meadow which covered nearly ten acres, everyone who could write signed his name and wrote also the name of the man next to him who could only cross it holding the pencil with his own hand. Most all had brought their rifles and a unanimous demand was made for someone who could drill them.

"My son will be here very shortly from the military school at Lexington," said an old cooper. "He has been there these three years past and will be mighty glad to help you in that line, for he has left the school in disgust, tired as he was to hear all their secesh nonsense and, for his bold stand, he was told by the commander that he had better go home. So he wrote to me that he had strapped all his things on his horse and would come home across the country, not the way Wise came down with his gang."

All promised to be there every afternoon and to bring along those of their neighbors whose names were not yet on the paper. It was also voted that, when it would have been signed by five thousand names, it would be mailed from Wheeling to the President of the Convention at Richmond or dispatched by a messenger if the mail could not be relied upon.

As soon as the news of this muster reached the ears of Wise, he ordered one of his companies to occupy the head of the road, back of the old refinery which led up the Elk, and gave minute instructions for a close watch upon all the movements of the force which he thought was organizing farther up the stream to come down and dislodge him from his position on the Kanawha. He also gave strict orders to blockade all travel on that road. Two days after, when he also heard that a young cadet had made his way from Lexington and was drilling the men by companies, he wrote at once for reinforcements to Richmond and to different points where, on his passage, he had started the enlisting of Rebel companies similar to his own.

At the close of the sixth muster, Mark announced that the paper had received the five thousand signatures, and a committee of five, of which he was made chairman, was appointed to secure a safe transportation of the memorial from Wheeling to Richmond. It was late in the evening when, after parting with Johnson, Towle, and Charles Lewis, he was describing at supper to his wife the incidents of the muster and unrolling before her the pasted sheets of the long memorial for which the other four members of the committee had agreed with him that Druker would be the most trusty carrier, he heard a low tap at the back door. Alice opened it. Her father, old Tom, taking off his cap to Mrs. Mark, made sign to her husband that he would like to speak to him if he would come to the door. Mark understood his ways and stepped out, shutting the door after him.

"What's the matter, Tom?" he inquired.

"Mr. Mark, my son John is down there by the creek behind the retorts. You know I told you once that he is digging coal back of Wheeling and has most bought himself out with his own money which he has earned by hard work. He has just come with a newspaper from Wheeling which tells how the Fort Sumter, near Charleston, has been shut down, and fired upon, and taken by the Secesh of Carolina. Would you like to see him?"

"Yes, by all means, Tom. Let us go down and see him."

As they went, he inquired again, "How did he happen to come so far away from his work?"

"Oh! That's not the first time, sir. He has been here before, and other boys too just like him. They keep us posted. They all know you however and what kind of a peaceful, good willing man you are. He brought one of these papers for you to read."

"Did you say his name is John, Tom?"

"Yes, sir, that's my name," said the young man who saw Mark when they had just reached a small thicket on the edge of the creek. He had been sitting on a log and held in his hands corn bread and bacon which he was eating like he was a hungry man in his hiding place. It was not dark enough to prevent Mark from being struck by his resemblance to his sister Alice, and the resolute, ready intelligence depicted upon his countenance; his frame was strong and well-knit, and the butt of a small revolver showed itself out of the breast inside pocket of his jean coat when he fumbled into it to pull off a folded newspaper tightly bound with a string to hand it to Mark.

"I told my father that I had brought this paper from Wheeling for you, Mr.

Mark. I knowed that old Wise had you here boxed up on the Elk, so that you could not get any more papers from Cincinnati to learn all what's going on and I thought I would come on my own hook through the woods, Wise or no Wise, and bring you some news."

"Many, many thanks to you, John," exclaimed Mark, grasping his hand and giving it a hearty shake. "When did you leave Wheeling?"

"On Monday evening, sir. This is my third night out. I got here Thursday afternoon. Major Anderson had to give up Fort Sumter after firing his guns more than twenty-four hours against all the guns around of the Secesh till his barracks took fire from their shells. This Wheeling paper will tell you all about it, also about the Convention which commenced there last week to make a new state out of this to be called West Virginia. I saw Mr. Freenberry in it, your good preacher from the Elk, and heard him, in his speech, say for the people on the Elk that they wanted to strike out for themselves, also the people on the Kanawha, and that those mush heads wanted nothing more to do with the mushy heads at Richmond and in old Virginiay."

As he was speaking, a few steps were heard on the other side of the thicket, and, turning found, Mark spied out, in the darkness, the forms of Alice and her mother peering through the branches.

"Here is your mother, John, and Alice too," said he.

"I have seen them already, sir, but I told them I was going back tonight and I reckon they want to see me again before I start."

"Yes, John," said his mother coming round the bushes, and holding a dark coat under one arm, Alice following her closely. "Give me this old coat of yours you have on and take this better one which Pole wants to give you in trade. I have filled all the pockets with bread and meat to last you till you get back to your coal bank. When will you be here again?"

"Pretty soon, mother." Then looking inquiringly at Mark, he added: "You have no powder, nor bullets, have you, Mr. Mark?"

"A little powder and small shot, John, and we have old lead to make bullets, but we need powder. We muster in the Elk meadow every afternoon and our men have been calling for ammunition." Pausing a while, he added: "John, would you take a letter from me to Mr. Freenberry at Wheeling? And bring back his answer? I will of course pay you for your time and trouble."

"I meant that very thing, Mr. Mark, when I just now asked you about powder and bullets. I knowed you wanted somehow to send word to Wheeling."

"I will be down in about an hour, John. Wait, will you?" said Mark.

"Yes, sir, I will," replied John.

"Shall I take him to the kitchen, Mr. Mark?" said Alice. "He can wait for you there till you are ready?"

"Yes, certainly," answered Mark. "Bring up also your father and your mother, that he may be with you all till he goes."

His wife was anxiously looking for his return. While waiting on her children, she had called Alice, and receiving no response had gone out to the kitchen and had not found her there. He told her the news which he had just heard from John and both heartily scanned over the Wheeling papers.

At five minutes of one o'clock, in the afternoon of the 13th, the garrison of Fort Sumter, after a continuous bombardment of thirty-four hours, had been compelled to capitulate, and to evacuate the ruins of the Fort, being allowed however the honor of war, to salute their flag, and take it out with them as well as their private and the company property. President Lincoln had called for seventy thousand men for three months, to suppress the insurrection against the Government, and had commanded all insurrectionary combinations to disperse within twenty days. He had also called an extra session of Congress to meet on the 4th of July. The news of the fall of Fort Sumter had aroused the people of the North to a just realization of their situation, and a general determination was expressed to support the authority of the Government. Excited meetings had been held in the various cities in the North. In New York and Philadelphia, great crowds had assembled and made demonstrations against suspected disloyal parties. Newspapers supposed to be tainted with secessionism had been compelled to hoist the American flag. The banks had advanced a great deal of money to their respective states for war purposes, and six hundred men of the 25th Pennsylvania militia had been sent from Philadelphia to Washington.

"Sarah, we will have to sew up a flag too and hoist it on the top of the works," said Mark, as he left the newspaper in her hands, and, seating at his desk, seized hold of a pen. "John is going back tonight and will take our memorial to Wheeling in a letter which I am going to write to Mr. Freenberry. He is a member of the Convention. This will save the trip to Druker and keep him to his family. His house is exposed on the Elk and needs his constant protection. John is willing to bring back an answer from Mr. Freenberry, and I will pay him for his trouble and time. Our mustered men clamor for powder and we must get it somehow.

"John is shrewd and determined. I will enclose this $10 gold piece to Mr. Freenberry and write to him to hire a horse on which John can come back and bring twenty-five pounds of rifle powder in two rubber bags carefully hidden in a long sack of corn meal. John can make the round trip in less than a week, and we will thus have regular communication with the outside world. I will also write a short letter to Col. Albert and request Mr. Freenberry to mail it at Wheeling for they must feel anxious about us. I will tell him the condition of affairs here and the course we have taken to protect ourselves."

His package to Mr. Freenberry, containing the two letters and the memorial, was soon carefully sewed inside of the back of one of his dark blue woolen shirts, and, telling his wife to go into their bedroom, he shut the door and called in the opportune messenger from the kitchen.

"John," said he, "here is in this small roll five small dollars in silver for your trouble, part of your pay. Put it away safely about you. Here is also one of my woolen shirts in the back of which I have sewed this small flat package of letters for Mr. Freenberry. I will get out a few moments and you will put it on. Call me when you have done. I write to Mr. Freenberry to hire a horse and send by you twenty-five pounds of good rifle powder in two small rubber bags hidden which you will hide in a sack of meal, you understand?"

"First rate, Mr. Mark."

"I tell Mr. Freenberry that you can make the round trip in less than a week."

"I can, sir."

"Now, I will leave you for a few moments. When you have changed shirts, knock at this door and I will come in. Your own shirt, you will give to your mother for your father but tell them and Alice also to say nothing about it to anybody."

"All right, sir," said John.

Within a few moments he had donned the new shirt as he had an hour before donned the new coat the pockets of which were as well filled with nourishing supplies as the back of the shirt was freighted with sound and effective patriotism and loyalty. He then knocked at the door.

"All ready now, John, are you?" said Mark as he came in followed by his wife.

"Yes, sir, all ready."

"This is my wife, John. She wants to shake hands with you for the good part you take in this matter, and also for the news you brought, bad as they are."

"I knowed you, Mrs. Mark, for I have seen you before from my mother's window down below."

"When was that?"

"Well since all this muss begun, I have come now and then from up from the Ohio to tell the boys what's going on and brought newspapers too. Some of them, Pole and Henry, can read purty well."

"What route do you take for these trips, John?" asked Mrs. Mark.

"Oh! I have several, ma'am, through the woods and up and down the creeks. In daytime generally I stay in shanties with colored folks. We have our talks, and I rest a while. On the horse however, when I come back, I will have to ride more in the open, but corn meal is only corn meal, and a good cover it is in a sack for me to sit on." Mark and his wife smiled, bade him good-bye and a safe return, and he went back to the kitchen. A few moments after they heard his hasty steps down the hill, and, when kneeling in prayer, they fervently implored the divine protection over him and the work which he had undertaken.

The next day, at the muster, Mark informed the members of his committee that, though he had not called on Druker to carry the memorial to Wheeling, thus leaving his little family unprotected on the very bank of a river open to any boat from its mouth, the paper was nevertheless well on its way to Wheeling and in safe hands, the owner of which he could not name to them just then, but he had brought to him, the night before, important news. There he laid before them by reading aloud from the Wheeling paper the fall of Sumter, the call for troops, the excitement through the North, and the proceedings of the Convention at Wheeling, the members of which were also named in the paper, and among them the well-known name of Mr. Freenberry, their preacher in the Elk valley, whom all knew so well. He added that they would all twenty shortly have a letter from him, and that twenty-five pounds of rifle powder would come with it.

At muster, when the long line had been formed, he suggested that it should form again ten men deep, to occupy less ground, and, standing upon a stump, with a loud voice, he repeated to the assembly his communication to the committee about the memorial, and also the intelligence printed in the Wheeling paper.

The surge which pressed around him upon his announcement, and while he was reading so that every word could be heard, was well calculated to strengthen his confidence in the determined attitude of his listeners, and, at his request which he based upon the momentous prospect of the civil war inaugurated by the fall of Sumter, Mr. Clark led all in a penitent prayer that the Lord, in His wrath over the enormities of a great national sin, would remember mercy, and bring to naught, in His own way and in His own time, the devices and machinations and malicious lead of a few daring, selfish men.

After Clark had pronounced the last word, Mark put it to the vote of all present to decide whether, in order to attend as they ought to their crops and other home duties, they would muster thereafter only once a week, say every Thursday afternoon, unless convened and called on extraordinary occasions for special purposes by the present Committee of Five. His suggestion was adopted by a large majority. Not a few of the young bloods however protested that they would come anyhow every afternoon for drill, and Mark assented to the idea as a good one if adhered to in an orderly way. The Lexington cadet gave his word that he would see to that, it being in his line, and he had already been recognized, young as he was, as a leader and a strict disciplinarian.

"In the meantime, Druker," said, Mark while walking away back toward the landing, "watch all pirogues and canoes on the Elk coming up from below to spy our movements. Have pickets every half mile down and let them light pine knots at night to be seen by every paddler sent up to spy in the shadow of the shores. All this will be reported below and will add to our strength. They will feel that we are on the lookout and on our guard, perhaps preparing a descent, who knows?" said he, smiling.

"We will, sir," replied Druker. "We will picket both sides of the Elk."

When they arrived at the landing, Mark looked down toward the small cove below the home of Druker.

"Both our boats are sunk in that cove, both safe upon the sand?" he inquired.

"Yes, sir, square upon it since I pulled off the three plugs from the bottom, and nobody can get them off. They know too that I am somewhat of a good shot, especially since I lost my right eye."

"Yes, by your gun bursting a few years ago while shooting one dark night at the bright eyes of that poor deer at a salt lick, wondering it was at the light of your tin lamp stuck upon your hat. Your wife told me all about it. Your light fell with you and the deer escaped."

"Yes, sir, but I hope those fools down there will know better than the deer. With what sense they have left, I hope they won't come to try me here."

"Well, good-bye again, Druker. I must ride back to the works. Tomorrow, I think, my wife will have our Union flag ready to hoist up into the wind. Towle erected the pole this forenoon upon the front gable of the roof over the retorts. If it's all ready, I will hoist it up tomorrow at sunrise after charging the retorts. Good-bye."

On his return he found the star-spangled banner all ready. His wife had invited to her aid the willing fingers of Mrs. Johnson, and both, in the afternoon, had been busily at work, with the children all joyous, sewing long pieces of red, white, and blue flannels around the thirty-five stars. At six in the morning, as the sun was rising above the hills and when the retorts had all been charged, Towle stood below the gable end of the tall building holding in his hands both ends of the long line which descended from the three small pulleys that bound it from the top to the middle of the staff, and, around him, stood Johnson, the carpenters, the coopers, and all the Black hands: miners, chargers, refiners, and firemen. Every eye was turned up toward the house on the hill in which Mark and his family lived, and soon he was seen coming out followed by his wife, his children, and Alice all smiling, as he was carrying the long folds of the flag, his little son Samuel holding the trailing ends from touching the ground. The father and the son thus came down, and they had walked but a few steps when Black Pole, inspired by the sight, threw up his old slouch hat, and, at the top of his voice, hurrahed, "Long live the old flag!"

"Hurrah!" responded fifty voices; "Hurrah!" repeated female and children's voices from the houses in the hills. "Hurrah!" cried also a group of men white and Black who were coming down on a run from the works above. "Long live the old flag!" hurrahed Pole again, and twice the same response greeted the Black man's cry for freedom. By this time Mark and his son had reached the line. The flag was immediately made fast to it, and when Mark, taking hold of the end, uncovered his head, all followed his example. The Doctor and his two brothers had also come, the Doctor's wife having hastened up the hill to the side of Mrs. Mark with her two children also. As the banner went up, Mark, following it with his eyes, impressively shouted "God save the old flag!"

"Hurrah! Hurrah! Hurrah!" cried every voice thrice again. A soft spring breeze was blowing and wafted by it. The national colors, in all their pristine beauty spread out a float to the joy of every eye, the glowing thought of every heart.

"I thank you, Mr. Mark, for the reading of this paper," said the Doctor handing to him the Wheeling paper, which, on his return from the muster, Mark had sent to him the evening before. "How did you get it?"

"I will tell you some other time, Doctor, but you see the madness of these leaders at the South and of their man Wise here at our Court House. Is it not our duty clearly to keep it from spreading ruin and death just as we would arrest the murderous antics of a lunatic running madly head long with a dagger in his hand and striking right and left?"

"It is indeed our duty, sir," replied the Doctor. "Wise will soon find himself at the end of his rope. Ohio, Indiana, Illinois will not tolerate him long where he is and will soon send him back up the Kanawha with all the reinforcements, which he has written for to Lexington and Richmond. I hear from the Licks that he talks very much like a disappointed man. He expected before he came that he could carry everything before him and would soon have an army to follow him down to the mouth of the river."

"The old fool!" exclaimed Towle who, with all the men, stood listening.

"Yes," added Johnson. "I reckon he knows by this time that he is a born one."

"I must go and thank Mrs. Johnson for her good help to my wife yesterday," said Mark to the Doctor, and as they were walking up the path, the latter inquired: "Are you well provisioned against this blockade, Mr. Mark?"

"Yes, Doctor. We looked for troubles ahead and provided fully for six months to come. Our large new store is full, from cellar to loft. How is yours?"

"Well, we can stand a long siege too, and if you happen to want anything that you may not have thought of, don't hesitate to call on us."

"I thank you, Doctor. We shall expect the same from you." Both their wives were on the porch waiting for them.

"Mrs. Lewis," said Mark, "will you not now complete the good work, and sing with my wife Francis Scott Key's ode to our *Star-Spangled Banner*?" As he spoke, he pointed to its waving folds, his arm extended, and his eyes kindled with patriotism.

"I will, sir," she promptly replied, taking Mrs. Mark by the hand and leading her to the piano. Her grave alto and Mrs. Mark's clear soprano voices soon rose in unison, and as these two noble American wives and mothers, in the fervor of devotion to the country of their birth, addressed themselves to extol the national emblem under which God had given them all what they had, all what they were. The melody, under the practiced touch of Mrs. Mark, was caught by the voices of both husbands, and, through the open windows and doors, it soared, but was heard also by the delighted ears that were listening below, spell-bound and intent. The listeners remained motionless till the last sound of the last stanza could be heard from that American home.

The sixth night was passing since John had left on his return to Wheeling, Mark's wife had retired after her children, and he was seated at his desk, entering on his books all the work of the day, when a gentle knock at the back door stopped his writing and called him to his feet. The thought that it was John at once crossed his mind. He raised the latch, and there stood the young man, his face grinning with delight. He carried a long, heavy sack of corn meal or flour upon one of his shoulders.

"John," said Mark, "true to your word! Less than a week's time on this trip!"

"Yes, sir, with this sack of meal and all there is inside."

"Let me help you take it down. Where is your horse?"

"Back of the same brush below on the creek, Mr. Mark. The papers and the powder and a letter from Mr. Freenberry, they are all in this sack. Can you let me off again in about two hours as you did last week? I want to eat some and feed my horse, a good beast that Mr. Freenberry hired for me."

Alice had heard his voice and noiselessly opened her door.

"Alice," said Mark, "he wants to go back in about two hours. Get his supper ready as soon as you can, and in one of the meal-sacks which you have in the kitchen you will pack his meals to last him a week. You stay with her, John, till I am ready for you. She will go down and tell your father and your mother that you are here, and he will let you have all the corn you want for your horse now and for your journey." Then, closing the latch again, he turned and saw that his wife had partly opened the door which led into their bedroom. She had been listening to his short conversation with John.

"He can be relied on like his father," said Mark opening the sack carefully over a large tub, and into it the meal began to fall as soon as he had loosened the string twine which tied up the end.

"Here is the powder in this long rubber bag, a ten or twelve double one too, well tied up, about twenty-five pounds in it. Our men at the muster will be glad to see it on Thursday."

"Here are the papers too!" exclaimed Mrs. Mark, as she helped him to hold the sack, "and a letter tied on the bundle but this tub is full, hold up now the empty sack which you laid on that shelf for this purpose."

"Here it is," replied Mark. And opening it, he continued emptying the rest of the meal in which he also found the other rubber bag filled with powder.

"Let us read the letter first," he said when the sack was emptied.

"From our good friend, Mr. Freenberry," he soon added looking at the signature after he had cut the letter opened.

"Wheeling, April 23d, 1861.

"My dear Mr. Mark.

"I thank you for your good letter by bearer John, and for the memorial. I have read it to our Convention which has ordered it printed and published also in our papers here that it may thus reach Richmond, all communication with the doomed place being now cut off. Governor Letcher has recognized Virginia as one of the Confederate states. He has ordered all U.S. post offices and other public property to be seized, the harbor of Norfolk to be obstructed, and Lieutenant Jones of the U.S. Army has been compelled to destroy the Government's armory and buildings at Harper's Ferry and all their contents, to prevent their falling into the hands of the Confederate force secretly sent by Letcher. Jones and his command marked thirty miles that same night to reach the U.S. barracks at Carlisle, Pennsylvania. Letcher has also offered $30,000 for the patent of the bullet mold, and the answer he has received is that 'no money can purchase it against the country.' Our Navy

> yard at Gosport, with all its contents, has also been destroyed by Commander McCauley to prevent their use by the Secessionists. The sloop of war Cumberland was saved by our Government. The loss to our Government is about fifty million dollars. Letcher has also appointed Robert E. Lee, who has resigned his commission in the U.S. Army, as commander of the Confederate military and naval forces of Virginia.
>
> "The newspapers which I send with this will inform you of the work of the devil these last few days in the other states of the South since he was permitted to contrive the evacuation of Fort Sumter on the 15th of this month. But the Judge of all the earth will do right. All our times are in His hands. I have mailed your letter to Col. Albert and written to him that you have secured a trusty messenger by whom his communications can safely reach you through me."

"A good thought," exclaimed Mark, "which came to me too late after John had gone, but Mr. Freenberry did not let it slip."

Then resuming the reading of the letter.

> "The powder I bought, fifty pounds, I bought with part of the $10 gold piece and will keep for what remains $3 the hire of the horse. Black John would go for you to the end of the world, if I may so express his feelings for you. You will find your memorial printed in these Wheeling papers. Remember me affectionately to all our people on the Elk and Mill Creek.
>
> Rev. Freenberry."

The sound of his voice, though lowered, having here disturbed the sleep of his little Sophie, her mother hastened to the side of her cradle, and a look at his watch admonishing him that the two hours would soon expire, he seized his pen and wrote a letter of thanks to Mr. Freenberry, enclosing another Eagle for another fifty pounds of powder and a renewed hire of the horse. In his letter he said: "Our men on the Elk grow more determined every day to resist secession, and Wise, I hear, feels most disappointed at the very small number of recruits he finds on the Kanawha for his Confederacy."

In another letter to Col. Albert, he wrote the same information regarding his corresponding with Mr. Freenberry through Maria's son, and his confident expectation soon to hear from Cincinnati and Frankfort through the same providential channel. He also gave as full a detail of the working of the mine, of the retorts, of the refinery, and of their accumulating stock in barrels and in tanks as time permitted him, and closed with the hope that Wise would soon feel thankful for a way of escape out of the Kanawha through Hawk's Nest at the mouth of Gauley, another tributary to the Elk.

Mark was a man of peace, but his mind was well balanced. He had keen perception and could see how easily this badly managed force might be turned and scattered, as composed of men spread out along the mountain-locked streams which the young Wise, in his boisterous way, had come to possess and to hold. Three months later he could in silence muse and smile at his hitherto unconscious possession of innate strategic capabilities.

After rolling up five silver dollars in a piece of paper for John, he opened the back door to go to the kitchen, and was not a little surprised to find him seated upon the step, silent and watching. Alice's door was opened, and upon a bench near it, she also was seated with her father and her mother. John's sack lay at his feet well filled.

"Come in, John," said he. John followed him, closed the latch, and stood hat in hand. "John," said Mark, "here is another five dollars for your week's good work. We want fifty more pounds of powder which Mr. Freenberry will buy again and send by you with more papers and another letter from him. Here is my letter to him. How will you carry it safely?"

"I'll pin it with three or four pins inside the back of my flannel shirt, Mr. Mark, as you did with the last. Please give me a small paper of pins."

"Here it is. Now we will look again for you a week hence."

"I'll be here, sir. God will take care of me. You and Mrs. Mark, and my father, my mother, my sister Alice, you all pray that I go and come back safe. Then I know I'll be safe."

"Good words, John," said Mrs. Mark. "God hears and grants such petitions as this for the safety of an honest worker like you."

Shaking hands with both, John took the letter, bade them feel quiet on his account, and went out in the dark. The loud and measured strokes of the clock in the refinery could be heard announcing twelve o'clock, and, opening one of the windows, Mark and his wife, attentively listening, could catch the sounds of the receding steps of John, and of his father, his mother, and Alice, going down the hill toward the thicket where his horse was tied and had been fed. Not long after the sounds also of the horse's hoofs on the path distinctly reached their ears from up the creek path until they gradually died away, mingled the cry in the distance with the song of the whip-poor-will, the low shriek of the owl, and the startled query of the cow bell.

The next day was Thursday, muster day. It was again numerously attended for expectations ran high along the Elk and among the hills that Mark would in some way have news from Wheeling. He called on all present to form a hollow square in two ranks, and, mounted on his stump, read the public intelligence brought by the letter of Mr. Freenberry and confirmed by the papers which he had sent. These stated, in addition to his letter, that Kentucky, divided in her allegiance, had agreed to remain neutral, and refused to furnish troops to the Federal Government, that a detachment of the 25th Pennsylvania militia, under Col. Cake, had entered Washington—the first volunteers for its defense—that Major Anderson had arrived in New York. That the 6th Massachusetts and the 7th Pennsylvania militia had had a fight with a mob in Baltimore while passing through the streets on their way to Washington in which three soldiers had been killed, eight wounded, and nine of the mob had been killed and three wounded. That President Lincoln had declared a blockade of the Southern ports. That General Patterson of Philadelphia had taken command of the Department of Washington. That Philadelphia and New York had each appropriated one million dollars for war purposes, equipping volunteers and supporting their families. That the 7th New York regiment had left that city for Washington with forty-eight rounds of ball cartridges. That the Baltimore mob had succeeded in checking railroad travel from the North by destroying various draws and bridges. That Southern merchants were repudiating their debts to the North, that the U.S. Arsenal at Liberty, Missouri, the Fayetteville Arsenal in North Carolina, and the Napoleon Arsenal in Arkansas had been seized by the Rebels, and the Branch mint at Charlotte, North Carolina, by Governor Ellis. And that the American flag had been publicly buried at Memphis, Tennessee.

"These, my friends, are all the main events which have occurred since we last met a week ago. At our next muster, we will hear what is now taking place, and will take place during the coming week. Let us all go to our homes, trusting in the protection of God, and that He will surely overrule all this evil for a better Union and the welfare of our country!"

"Let us also keep our powder dry," shouted Druker.

"We will sure!" echoed more than four hundred determined, strong voices.

Eight weeks thus passed of devotion to duty shared alike by Mr. Freenberry at Wheeling, by Black John through the woods to and from Wheeling, by Mark on Mill Creek, and by the muster men on the Elk until Wise, having received his reinforcements and heard of the steady, if slow, approach of an Ohio brigade from the mouth of the Kanawha under Gen. Cox, took the advice of the Rebel friends who had called him from Richmond. They were well acquainted with the topography and told him that one of the Ohio regiments might cross over the Elk and join the Unionists there to fall upon and surprise him some night while he might be looking for the main force coming up the Kanawha, and that he would do well therefore to send up a strong body of his men to occupy and watch both banks of the Elk beyond the landing at the mouth of Mill Creek.

"Whom shall I choose for this command?" inquired Henry Wise.

"That young West Pointer who has just arrived from Kentucky with his Major's commission, Crone," they replied. "He seems a bold fellow."

"Yes, he is," said he, "but he drinks. His men might not be safe in his hands, and those Elk squatters are no mean foe, a resolute set."

"Oh, some of us will go with him and guide. We will watch and keep him straight."

"Well," answered Wise, "he has had military training, and comes of good stock: his father, Gen. Crone, did well in his day at Fort Pulaski in Georgia. I will send him with two companies and give him his instructions. The Lexington cavalry troop which came in yesterday will go with him. Fine young men, all of them, well mounted and well armed."

Late in the night of this same day, this conversation was reported almost verbatim to Mark by Pole who brought up to the house his Black informant, a friend, servant of one of Wise's advisers. His wife had overheard most of it in snatches at the supper table while appearing utterly unconcerned, and he had walked at once through the hills in the dark to report it to Mark and warn him and his men to be on the lookout.

"I thank you, Isaac, for coming so far and so late to put me and all of us here on our guard. You have not had much supper, have you?"

"No, sir, I did not feel hungry then. I was too much in a hurry, and I wanted to gain time."

"Well, let us go to Alice. She is in bed, but we will wake her up that she may give you a good supper."

"She is up at our window talking to me," said his wife, pushing the door and coming into the room. "She had heard Pole and Isaac knocking for you at this back door and was soon on her feet listening."

"All right. Tell her to cook some supper for Isaac. He wants to go back at once and reach home where his wife is waiting anxiously for him, before anyone else knows that he has been this way."

"Thank you, Mr. Mark. I feel a little hungry now and eating some will help me to walk back."

"I'll go part of the way with him," said Pole, "to show him a shorter road than that he came by. 'Twill save him nearly three miles, and he won't have no pickets to bother him."

"Two hollered 'Halt!' to me," said Isaac, "but I laid me down low like a possum and said nothing till they thought, I s'pose, they had only heard a cow in the bush and turned their backs. I then crept on as soft as I could."

"He came by Clarkson's hill, back of Elk, that's the reason," said Pole to Mark, "he did not know there is a higher way by the sugar grove and no pickets that way."

"That's right, Pole. Go with him till you see him perfectly safe. He has been a friend to us tonight."

"I'll come and see you again, Mr. Mark, and keep you posted."

"Do so, Isaac, but be careful, very careful. Do not risk your life. You came very near being shot at tonight."

"I did not know enough of the right way, but I'll know better next time. I'll be careful, Mr. Mark."

Pole opened the door, Isaac following him, and, as they stepped out, Alice stood at her own door, having lighted her lamp. She beckoned to them.

"Come on, both of you," she said. "Mr. Mark wants me to give you some supper before you go and 'twill soon be ready. Sit down, Isaac. How is Lizzie?" she inquired, referring to his wife whom she knew well.

Thus conversing with them, she soon gathered upon the table a fragrant supper of fried ham, eggs, and corn bread with butter and milk. Both did full justice to it and, a few moments after midnight, through the darkness, the firemen at the retorts could descry the forms of the two men, whose errand they knew, across the creek, climbing the hill on the other side. All knew the report which had been brought by Isaac. All were discussing at low voice, as they met in their rounds, what was going to happen on the Elk, but, for all that, each one kept his fires bright and his grates clear.

Mark was among them in the morning, and testing the run of a still when Druker, arriving from the landing, approached him, evidently trying hard to suppress some strong emotions, for, when Mark greeted him in his usual quiet and cordial manner, he could hardly utter a word.

"What is the matter, Druker?" asked Mark looking at him while holding in his hands a hydrometer and a glass test tube.

"Can you come up to your house, Mr. Mark, for a few moments?" answered Druker. "I have some news for you."

"Yes, let us walk up. I suspect however what it is, Druker."

"Have you heard already of the mounted Rebel devils who came up the Elk last night?"

"I heard last evening that Wise was going to send them to watch the other side of the Elk. He fears the crossing over there of one of the Ohio regiment, which Gen. Cox is bringing up along Kanawha."

"I heard the tramp of their horses, Mr. Mark, just after midnight. I jumped out of bed, my wife after me, and, from our window, through the dark we could see them all passing up the road in front of our orchard two by two, my dogs barking at them at the fence mad as fury and still barking after all were gone. As they continued barking, at about two o'clock, listening close, we thought they were after someone on the bank of the Elk among the elms just above our house, some picket, we thought, which might have been posted to watch our movements, and, from the loft under the roof I could hear, at about a mile or so away, around the big barn of that rebel Chap, above us a piece, as if those mounted fellows had stopped and tied their horses, perhaps to feed them. 'Good for Chap,' I told to my wife, his and them will go tonight sure and in a hurry to feed all these beasts of his good friends, but nary a dollar will he get from them, and who knows but, when they go, they'll take him along too, for his company's sake, on one of his horses and with his rifle strapped on his shoulders just like one of themselves. I had dressed and felt much like going out on the road to see and take observations, but my wife would not hear of it. 'They will take you too,' she said, and 'force you along with them. Keep quiet here till daylight, and, when I go out to milk Sucky, if there is, as I think, a man posted on the bank, I will go and speak to him and make him talk. I will too have from him all we want to know about his band of night prowlers.' As the light of morning came, kneeling on the floor of our loft, I could see, through a crack between the logs of our house, that there stood a man with a gun leaning against one of our tall elms and facing the other side of the river, the trunk of the tree partly hid him from us. At about four o'clock, our dogs still barking with a will, four or five armed Rebels came down the Elk along the bank. The picket stood up straight with his gun on the ready and his finger on the trigger, and halted them, challenged them too as if they were fresh from Ohio.

"The one in command came toward him saying, 'Friend!' They spoke. At a signal from him, the rest came down also, and soon all went back, leaving another man on guard. Surely then we were watched as well as the other side of the river. But you know my wife. She is gentle as a lamb, but she is fearless too, and would face a panther to keep it from springing on her children. Sure enough, at about five o'clock, while I was in the loft yet, she went out with her pail, calling for Sucky and toward the man among the trees. I brought my rifle to a ready. The man was standing looking at her, and his gun which he had grabbed at first, he seemed to forget it, for he let it down to a rest, neither did he halt her.

" 'Who are you?' said she. I could hear her. 'What are you doing here with your gun?' The dogs, as she called them, Murky and Stumpy, stopped their barking, standing at her side however and looking up fiercely at the man and his gun.

" 'Who are you, yourself?' he answered, but his voice had not harshness in it.

"My wife kept on toward him in her gentle way, you know as she trips along, with her milk pail in her right hand, and Sucky behind with her calf approaching behind her too.

" 'How you remind me of my own Suzy in the morning!' I heard him again. 'You look too just like her.'

" 'Your wife?' said Mary.

" 'Yes.'

" 'Where do you both live? But you did not tell me what you are doing here with your gun so near our house?'

" 'Oh! I wish I was home again in Rockbridge, and not here on this nasty business, trying to kidnap good men from their wives and children!'

" 'Is that what you are here for?'

" 'Yes, but the first chance I get, I am off, you may be sure. I have enough of their yarns. Milk your cow, my good woman, and please let me have some of your milk in this canteen. I have not much money, but I will pay on for it.'

" 'For your wife's sake, I won't take your money, sir, but I will fill your canteen, just as she would fill my husband's, were he as far from home as you.'

"I saw tears start from his eyes when she stooped to milk Sucky. Both of our children had followed her carrying between them, in another pail, the cow's warm mush for her breakfast.

" 'Your husband is in the house, is he not?'

" 'Yes, sir, why do you ask?'

" 'Well, I will tell you. As I rode last night not far off from our Major who commands our troop, I could hear him half drunk and half sober, speak to some gentlemen who seem to be with him as guides through this country, and he was disputing with them about a raid which he is decided on up a creek, Mill Creek, he called it, and they were trying to persuade him against it. They said he and all his men would surely be shot down like sheep if he attempted it. But he insisted upon it, that he would make it, even though Gen. Wise had not told him to do it, as his name was Crone. He would certainly go up to that Yank Mark, as he called him, in his lair up that creek, this very night coming, and capture him dead or alive or else make a bonfire of the coal oil works which he runs there. He seemed to know all about him, his wife, his children, all the white and Black men who work for him, that he is in the employ of some rich men in Kentucky who are on the Union side and doing their level best to keep their state in the Union. Do you know this man Mark?'

" 'I do, sir. Give me your canteen,' and she told to one of the children to run to the house for a small pitcher to fill the canteen with. 'Yes, sir,' she continued, 'I know this Mr. Mark, and his good wife, and his fine children, and all about him. And you had better keep off your hands from trying to touch one hair of his head. I will certainly send word to him to be on the watch against this raid. Are you going too?'

" 'Not if I can help it. I have not enlisted to kidnap men in their beds at night and on the sly, as this Major intends to do. It's a cowardly trick.'

" 'Well then, if I were you, the river can be crossed here. It is very shallow, I would go right up on the other side, and make my way to the headwaters to go back to your home in Rockbridge, and say to your wife and children, 'You need me and I need you. I will not be deceived again to leave you unprotected.' Here is your canteen, sir, and the Lord go with you to guard you.' "

They were seated in the house and Mrs. Mark was listening.

"You have a noble wife, Mr. Druker."

"She is, madam, and I was not slow, I assure you, as soon as she repeated all this conversation to me, to think as she did, and I came at once to tell it all to you both."

"This Major Crone, then," inquired Mark, "was half drunk and half sober when he spoke to his friends and they were opposing him, calling his design foolhardy? They were right, and my opinion is that, by the time he thinks as they do, and, as long as they can keep whisky from him, he will follow their advice, staying where he is now, watching the other side of the Elk for that Ohio detachment which may be coming to join us, although so far no General word has reached us yet from General Cox."

"Chap has plenty of whisky, Mr. Mark, and we'll treat that Major."

"Yes," said Mrs. Mark, "and those gentlemen who are with him will not be able to restrain him, unless they have time to send word to Wise and get his order prohibiting such an attempt as this." An anxious look spread over her face as she addressed this warning to her husband.

"Well," Mark replied, "whatever wrong God permits, He overrules also, for the good of all whose whole trust is in him. I am ready for whatever happens. Had you not better stay here with us, Druker, till that Rebel troop goes back? They cannot stay very long there."

"This is what my wife told me this morning, but," he added hesitatingly, while looking at his rifle, "I don't like the idea of paying so much consideration to such a rabble as that, and leave my wife and the children to their tender mercies."

"Oh! She has proved to you early this morning," said Mrs. Mark, "that she is more than a match for any of them."

"That's true, madam," Druker said with exultation in his eye. "She can reel them round her fingers and they won't know it."

"You stay with us, Druker," said Mark.

"And I'll go and tell Mrs. Druker," put in Samuel, who, with his sister, had come in while this conversation was going on and had been eagerly listening to it.

"Yes," added Mrs. Mark, smiling. "They can go now and tell your wife. They can keep on the ridge till they come to the road on the river and run across it to your house, then come back as quick as they went." So speaking, she was looking down both at Samuel and Hattie.

"Yes, we will," said both children, clapping their hands at the idea of being allowed to take a part in the war.

"Did you see any pickets, Druker, as you came up?" inquired Mark.

"No, sir. I came also by the ridge, but I would not be surprised if some were posted along the creek, a short distance up this way from the landing."

The children went, delivered their messages and returned before noon, elated with their success. They had seen Mrs. Druker, and she was glad that her husband had assented to follow her advice.

"She could take care of herself and of the children," she had said and, to prove her confidence in her surroundings, she had sent her two little ones to keep company with Samuel and Hattie through the thick timber upon the ridge part of the way. Peering down from the cliffs toward the creek, they had seen three men in three different places, each with a gun, the last at about a quarter of a mile from the landing, and looking "scary" as Samuel termed it, "looking around all the time" was Hattie's report.

"They are afraid of us here," said Mark with a smile. "They know well that but a few bold men, from the timber and rocks on each side of this creek, which we all know so well, every foot of it, could shoot down every one of theirs in a raid upon us here."

Resting in this confidence that Crone, even if maddened by drink, would be held in check and kept on the Elk, Mark made no preparations to resist and went all day about his duties in the works. When night came, after supper, and family worship, and his wife and children had retired to bed, he continued making all entries for every department of the day's work.

At about eleven o'clock, he had just closed his accounts and laid down his pen to retire, when his ear caught, through the stillness the night, at times, an unusual noise coming up from the lower bends of the creek. He sat listening. In a moment his wife was at his side. She also had been wakeful and had heard the same unusual sound.

"This is a tramp of horses," she said.

"I think it is," he replied looking at her. "Can it be possible that he would have the hardihood to undertake such a venture as this?" he continued, musing to himself.

The noise became more distinct.

"You have no time to lose," she hurriedly said. "Throw your gray shawl over your shoulders and run up the hill to the ridge among the trees before they can see you."

As she spoke she took it down from the hook on which it hung, and, opening it, threw it over his shoulders and opened the back door. He had made but a few steps out when hoarse voices from outside the picket fence which encircled his garden cried out from three different points: "Halt! Halt! Halt!"

"Who are you?" he answered as, jumping over the fence, they advanced toward him, each from his post, their guns ready to fire, and looking steadily at him.

"Never mind who we are," replied one of them. "You are our prisoner, and don't fuss. You come down with us."

He could, through the darkness, observe, at the foot of the hill, a large number of mounted men standing still in the road.

"Is Major Crone down there with you?" he inquired from one of his captors as they were taking him down.

"Don't ask any questions, the less you say the better for you."

Resigned, he walked with them, in his mind quietly looking up to God in whose hands were the hearts of all these men, and musing on this, he thought: you can only do what He permits you to do, and you shall soon see it to pour out confusion and shame, every one of you, poor deluded men.

"Is your name Mark?" asked an officer on horseback at the head of the columns as soon as they reached it.

"My name is Mark, sir, and what is your name?"

"Never mind my name. You are my prisoner, and stay here quiet till I come back."

Then he rode on quickly, up to the house, one of the men opening the gate for him.

"Has your husband any arms here in this house, a pistol, a gun?" he hurriedly inquired from Mrs. Mark who stood in the porch with Alice and the two children.

"My husband has no arms, sir. It is well for you that he has none."

"Better for him, madam."

And with a bow which he had learned at West Point, he spurred his horse back to his men. His voice and his manners were excited.

"What are you going to do with Papa?" cried Samuel to him. "Are you a band of robbers to come at night so?"

"Yes, and steal him away," added his sister.

"Shame on you," continued Alice, with as loud a voice as she could raise. "Ye are no gentlemen, but a band of kidnappers, every one of you."

Under the guard of the three men who had brought him down, Mark stood quiet and composed, waiting for what disposal was to be made of him. He could see, from where he was, none of his men about the works. He knew that not a few of them were posted behind trees or rocks upon the hills around with their rifles in their hands, and that, if he but gave the order, they would open fire and what would happen? This was the question in his mind: a panic might seize his captors and set them in a run down to the creek, but their leader was evidently bent on mischief regardless of personal risks, and if left alone or with a few, might he not carry out his threat, and, in his madness, set fire to one of the many large wooden oil tanks which lay before him, after knocking the light sheet iron lids which covered each of the retorts.

October 29, 1887
This is only the beginning of the manuscript the 4th part; before the 1st of December I will endeavor to mail the remaining three fourths which have to be copied as this fourth part has been. T.M.

[Editor's Note: The rest of the manuscript has been lost, though Theophile maintained he would copy the other sections and send them to his son. In 1861, Theophile himself was captured briefly, and in this manner. See the appendices to learn more of Theophile's own brief captivity and life during the war. He later enlisted in the Union Army, but, due to his age, was assigned engineering and administrative work far from the front lines.]

APPENDIX A

"MADE UNION FLAG: A STORY OF THE CIVIL WAR"

The Evanston Press, JUNE 1, 1901.
MADE UNION FLAG: A STORY OF THE CIVIL WAR
MR. AND MRS. MAHER OF WILMETTE HAVE HAD THRILLING HISTORY

WHEN MR. MAHER WAS IN FRANCE.

If you should go to Wilmette, and you will if you read the story they told me there, seek out a quaint green cottage among the trees where Mr. and Mrs. Theophile Maher live, and listen to them tell the stirring story of their lives. They have been a part of some of the most interesting history of the century just closed. In 1860, when the fires of secession and rebellion were blazing fiercely in the south, Mr. and Mrs. Maher were living on Mill Creek in West Virginia. Mr. Maher had gone there to take charge of some large coal oil works. The oil well of the present day had not then been discovered, and coal oil was made by crushing cannel coal and then extracting the oil from it. The debate waxed warm between union and non-union men, and time for voting upon the question of separating from the eastern part of the state and staying in the Union was to be voted upon. There was no American flag in the neighborhood, and one was much needed to rouse the enthusiasm of the voters and make them vote for the Union.

HOW THE FLAG WAS MADE

Mrs. Maher was equal to the occasion. She told her husband if he would go over the mountains to Charleston, ten miles away, and procure the bunting she would make the flag. The materials were procured late Saturday and the election was to be held Monday. Steadily her fingers flew, and rapidly the flag grew, but the hours slipped by and the holy Sunday morning dawned; but Mrs. Maher sewed on till star and stripe were all complete and the American flag was ready to unfurl to the halting citizens. Mrs. Maher says it was the first and only time she ever sewed on a Sunday.

On Monday evening Mr. Maher and a few union men went to the polls, and unfurling the flag, called upon the sturdy mountaineers to rally to its support as their fathers had done in the past. The appeal won and the election was carried to stay in the union. The flag was put up over the oil works, but as the war progressed Mrs. Vance, a rebel woman, got possession of the flag and tore it to bits.

The separation in families in the border state was illustrated by this same Mrs. Vance and her mother Mrs. Vicker. Mrs. Vicker was as enthusiastic for the union as Mrs. Vance was a rebel. Mrs. Vicker was in the house with her small children alone, all the union men having been forced to hide in the mountain, when a rebel fragment party came to her door. She stood in it with an ax uplifted and threatened to brain any man who crossed her threshold, and no man entered.

Rebels Seek Mr. Maher

News was brought to Mr. Maher that the rebels were coming to take him prisoner. In speaking of it he said: "I did not believe they would dare attempt to do it, but I told my wife if I was to be taken prisoner I had better get my book arranged so the company would not be interrupted in their business. I was working on the books about 11 O'clock in the evening when my wife came and said to me: 'The rebs are coming; you had best go out the back door and go to the mountains,' but I thought it would disturb some ladies who had taken refuge with us and were asleep in the room I had to pass through, so I went out the front door and walked into a rebel squad [and was taken to Charleston].

"My captor was Captain Crone, whose father so bravely defended Savannah in 1812. He was just examining me when someone came and whispered to Crone. He jumped up, remarking to me, 'you can go,' he hurriedly left the room. Gen. Cox of Ohio had arrived in time to rescue me."

Call for Recruits

On July 25, 1862, Mr. Maher issued the following circular: "Volunteers wanted. Men of Kanawha county, you have heard the call of the president for 300,000 additional volunteers, and the response of the governor for two regiments. How many companies will this county furnish? I want to form one. Who of us able to go can remain any longer a quiet spectator of our struggle for life and freedom? The offer to every volunteer is a liberal one. Upon his company being mustered in, he is entitled to one month's pay in advance, also $25 of the $100 bounty, making together $39 in advance. The allowance for clothing is $3.50 per month. He is also, or his heirs, entitled to the benefit of the invalid pension should he be disabled, killed, or have died in the service. Come and give your name and let us have the muster compete in the next two weeks. T. Maher, Charleston, July 25, 1862"

Mr. Maher has preserved one of these circulars, and his old face kindles as he proudly states his company was full in four days. Mrs. Maher and her children remained alone in her home. Mr. George Maher, the architect of Mayor Patten's new house, was one of the children left in the home when Mr. Maher enlisted. Mr. Maher is of French descent, his father having been born in the island of Santo Domingo. The latter was a young man at the time of Touissant L'Overture's [*sic*] rebellion, and was hidden in a molasses cask, and so escaped the fate of most of the whites on the island. He came to Philadelphia and there met Sophia Clavier of Nantes, France, and married her, and established himself in business at Baltimore. A portrait of him, painted by [Gilbert] Stuart, is in possession of Mr. Maher.

Knew the Elder Dumas

Mr. Maher was sent for his education to Paris and knew the elder Dumas. He was one of the enthusiastic crowd of students who drew Victor Hugo through the streets of Paris in triumph when his first drama was acted on the stage. The founders of the celebrated Paris Newspaper, "Figaro" were collegemates of Mr. Maher.

Space fails to tell one-half the interesting things Mr. Maher has seen. Among the relics cherished by his daughter is a watch of Mr. Maher's mother, a charming piece of artistic work in the eighteenth century. The picture on the back is delicate piece of repouse [*sic*], made by using gold of different shades and silver. The figures are raised and of exquisite grace and beauty.

These charming old people have seen so much and tell it with such picturesqueness, and make the caller so much at home, it is almost like living through the scenes they describe. The evening of their lives is passing pleasantly. A daughter, Miss Maher, is still in the home with them, and their children are near enough to visit them.

Appendix B

"The Flag That Saved a County to the Union"

Pages 138–39 in *Our Nation's Flag* (1903)

The Death, on the eleventh of March 1903, of Mrs. Sarah Landis of Wilmette, Illinois, calls to mind an incident of peculiar interest associated with the beginning of the Civil War. The Chicago *Record-Herald* says that Mrs. Maher and her husband were then living in Mill Creek, Kanawha County, which is now a part of West Virginia, where Mr. Maher had charge of coal and oil works. The sentiment was running strongly to the side of division of the Union, when Mrs. Maher urged that, as there was not a Union flag in the district, her husband ride to Charleston, ten miles away, and procure silk materials so that she could manufacture one. The question of secession from the Union or from the eastern part of the State was to be voted on two days after Mrs. Maher received her silk, and she was compelled to work all day Sunday and all Sunday night to get the flag finished in time. The next morning, when the men of the district went to the polling place, they found floating above it a beautiful banner, Mr. Maher on a block ready to address them, and his wife by his side pointing to the flag as its folds flapped from the staff. Mr. Maher made an impassioned speech, and with tears streaming down his cheeks, begged his neighbors to remain true to the Union. The appeal was not made in vain, as a large majority was given against secession. Mrs. Maher then took the flag and, aided by a large body of men, planted it on a hill overlooking the whole valley, where it stayed till it fell in tatters.

Appendix C

Obituary of Sarah Landis Maher

Sarah Landis Maher: Heroine of the Civil War Passes Away at Wilmette

Mrs. Sarah Landis Maher of Wilmette, heroine of the civil war, died yesterday at the age of 73. She was the daughter of Abraham Landis of Lancaster County, Pennsylvania, the first prominent member of the famous Landis family of that state. Her husband, who died last December at Wilmette, at the age of 87, was a construction chemist, and erected the first sulphuric acid plant in the United States in Philadelphia in 1842. He also erected plants in California and Chicago in the early '50s.

Mr. and Mrs. Maher helped to make history at the time of the Civil War, for their efforts saved Kanawha County, West Virginia, to the Union. They were living at the time at Mill Creek, where Mr. Maher had charge of coal oil works. The sentiment was running strongly to the side of the secessionists when Mrs. Maher urged that, as there was not a Union flag in the district, her husband ride to Charleston, ten miles away, and procure silk materials so that she could manufacture one. The question of secession from the Union or from the eastern part of the state was to be voted on two days after Mrs. Maher received her silk, and she was compelled to work all day Sunday and all Sunday night to get the flag finished in time. The next morning, when the men of the district went to the polling place, they found floating above it a beautiful banner, Mr. Maher on a block ready to address them, and his wife by his side pointing to the flag as its folds flapped from the staff. Mr. Maher made an impassioned speech, and with the tears streaming down his cheeks, begged his neighbors to remain true to the Union. The appeal was not made in vain, and a large majority was given against secession. Mrs. Maher the took the flag and, aided by a large body of men, planted it on a hill overlooking the whole valley, where it stayed until it fell into tatters. Mrs. Maher was for many years a leading member of the Presbyterian Church in this city. The burial will be at Rosehill.

[Editor's Note: This obituary was found in Maher family papers. No other documentation is available.]

Bibliography

Abein, Paul. *William Dean Howells and the Ends of Realism*. New York: Routledge, 2005.

Abzug, Robert H. *Cosmos Crumbing: American Reform and the Religious Imagination*. New York: Oxford University Press, 1994.

Atkinson, George Wesley. *History of Kanawha County from Its Organization until the Present Time*. Charleston, WV: Office of *The West Virginia Journal*, 1876.

Birkbeck, Morris. *Notes on a Journey in America*. London: Severn, 1817.

Booth, Stephanie Elise. "The American Coal Mining Novel: A Century of Development." *Illinois Historical Journal* 81.2 (Summer 1988): 125–40.

Cash, W. J. *The Mind of the South*. New York: Knopf, 1941.

Davis, Rebecca Harding. "Life in the Iron Mills." 1861. Reprinted in *A Rebecca Harding Davis Reader: "Life in the Iron Mills," Selected Fiction, and Essays*, edited by Jean Pfaelzer, 3–34. Pittsburgh: University of Pittsburgh Press, 1995.

Dorsey, Christopher. *Southern West Virginia and the Struggle for Modernity*. New York: McFarland, 2011.

Enri, Henri. *Coal Oil and Petroleum: Their Origin, History, Geology, and Chemistry*. Philadelphia: Carey, 1865.

Gesner, Abraham. *A Practical Treatise on Coal, Petroleum, and Other Distilled Oils*. 2nd edition. New York: Bailliere Brothers, 1865.

Goudie, Sean X. *Creole America: The West Indies and the Formation of Literature and Culture in the Early Republic*. Philadelphia: University of Pennsylvania Press, 2006.

Graham, David Alan. *A Treatise on the Comparative Commercial Values of Coals and Cannels*. London: Spon, 1882.

Hambleton, James Pickney. *A Biographical Sketch of Henry Wise*. Richmond: Randolph, 1856.

Hatch, Nathan. *The Democratization of American Christianity*. New Haven: Yale University Press, 1989.

Justus, James H. *Fetching the Old Southwest: Humorous Writing from Longstreet to Twain*. Columbia: University of Missouri Press, 2004.

Lewis, Ronald L. *Black Coal Miners in America: Race, Class, and Community Conflict*. Lexington: University of Kentucky Press, 1987.

Owens, Richard H. *Rogue State: The Unconstitutional Process of Establishing West Virginia Statehood*. New York: University Press of America, 2013.

Rice, Otis K. and Stephen W. Brown. *West Virginia: A History*. 3rd edition. Lexington: University of Kentucky Press, 2010.

Smith, Nicholas. *Our Nation's Flag in History and Incident*. Milwaukee: Young Churchman Press, 1903.

Smith-Rosenberg, Carroll. *Disorderly Conduct: Visions of Gender in Victorian America*. New York: Oxford University Press, 1985.

Snell, Mark A. *West Virginia and the Civil War: Mountaineers Are Always Free*. Charleston, SC: History Press, 2011.

Stowe, Harriet Beecher. *The Key to Uncle Tom's Cabin: The Original Facts and Documents upon Which the Story Is Founded*. London: Sampson, 1853.

Venable, William H. *The Beginnings of Literary Culture in the Ohio Valley*. Cincinnati: Clarke, 1891.

Watts, Edward. *An American Colony: Regionalism and the Roots of American Culture*. Athens: Ohio University Press, 2002.

www.ingramcontent.com/pod-product-compliance
Lightning Source LLC
LaVergne TN
LVHW091147080826
845145LV00008B/2289

* 9 7 8 1 9 5 2 2 7 1 1 2 0 *